I0822985

## ALSO BY DAVID ARIOSTO

*This Is Cuba:*
*An American Journalist Under Castro's Shadow*

• OPEN SPACE •

# ·OPEN SPACE·

## FROM EARTH TO ETERNITY— THE GLOBAL RACE TO EXPLORE AND CONQUER THE COSMOS

DAVID ARIOSTO

ALFRED A. KNOPF
NEW YORK 2026

A BORZOI BOOK
FIRST HARDCOVER EDITION PUBLISHED BY ALFRED A. KNOPF 2026



Published by Alfred A. Knopf, a division of Penguin Random House LLC,
1745 Broadway, New York, NY 10019.

Knopf, Borzoi Books, and the colophon are registered trademarks of
Penguin Random House LLC.

Portions of this work were previously published in
*Aerospace America, Noema,* and *SpaceNews.*

Unless otherwise noted, all photographs are from the author's collection.
Intuitive Machines: page 28, page 31 (top and bottom), page 151, page 156,
page 162 (top), insert page 6 (top), insert page 7. NASA: insert page 5.
Alamy (NASA / piemags / Alamy): insert page 4.

Library of Congress Cataloging-in-Publication Data
Names: Ariosto, David, author.
Title: Open space : from earth to eternity—the global race to explore
and conquer the cosmos / by David Ariosto.
Description: New York : Alfred A. Knopf, 2026. |
Includes bibliographical references and index.
Identifiers: LCCN 2025025399 (print) | LCCN 2025025400 (ebook) |
ISBN 9780593535035 (hardcover) ISBN 9780593535042 (ebook)
Subjects: LCSH: United States. National Aeronautics and Space Administration |
SpaceX (firm) | Space tourism—Technological innovations | Space tourism—Moral
and ethical aspects | Space race—Economic aspects | Space race—Moral and
ethical aspects | Space flight—Technological innovations | Outer space—
Civilian use—Economic aspects
Classification: LCC TL794.7 .A75 2026 (print) | LCC TL794.7 (ebook)
LC record available at https://lccn.loc.gov/2025025399
LC ebook record available at https://lccn.loc.gov/2025025400

penguinrandomhouse.com | aaknopf.com

Printed in the United States of America
1st Printing

The authorized representative in the EU for product safety and compliance is
Penguin Random House Ireland, Morrison Chambers, 32 Nassau Street,
Dublin D02 YH68, Ireland, https://eu-contact.penguin.ie.

*To my mother,*

*who instilled in me a sense of wonder in the cosmos.*

*To my daughter,*

*to whom I hope to impart the same gift.*

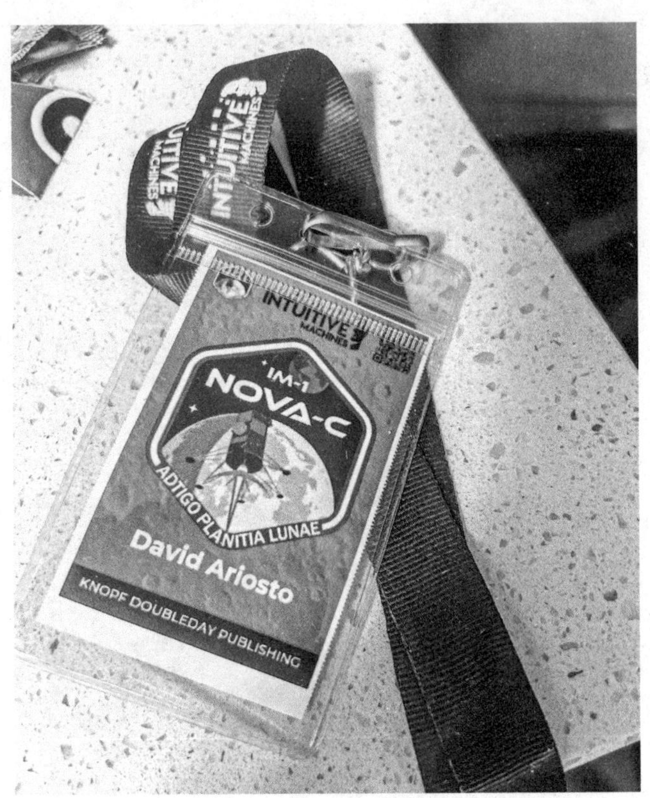
INTUITIVE MACHINES
INTUITIVE MACHINES
INTUITIVE MACHINES
IM-1
NOVA-C
ADTIGO PLANITIA LUNAE
David Ariosto
KNOPF DOUBLEDAY PUBLISHING

# Contents

• OPEN SPACE •

# PROLOGUE

Go to the wild places.

Travel to where only pinpricks of starlight interrupt the blackness of night, where nature's raw struggle is more evident. Then look up, as our ancestors did, and tell stories. Unlike those of the ancients, who sought to explain the cosmos without physics or modern astronomy, today's space stories tend to be rooted in the work of engineers, scientists, and entrepreneurs. Yet, curiously, their accounts are no less wondrous than the legends that they've replaced. Rockets the size of skyscrapers now land upright. Astronomers stare back in time to a moment just after time itself might not have existed. And physicists conjure new forms of energy out of what appears to be sheer nothingness. There seems almost no limit to what we might achieve.

In the course of writing this book, having by now held in my hand a device that magnetically suspends antimatter—the twin of almost all subatomic particles in the universe—I have come to believe that humans are capable of engineering just about anything. Yet even as we devise the kinds of technologies that unlock new, almost Copernican-level understandings of our place in the universe, the headwinds at home have grown fierce, just as competition from the East is rising. In the United States, the funding of basic science not only is shrinking but also remains increasingly hostage to political agendas, just as rivals—notably Beijing—treat the cosmos for what it is: a burgeoning arena of power, discovery, and economic interest. Space, which has long sparked marvel, has never before held so much potential, demanded so much stewardship, or risked so much in its abandonment. This is not merely about rockets and spacecraft but instead reflective of a broader evolutionary contest over who builds the systems, sets the rules, shapes the markets, and ultimately claims the rewards in this next era of human advancement.

Perhaps, however, we should start with a more elemental premise, a question that has plagued all manner of space explorer throughout the ages: Why should humans chase the stars at all? There are several ways to answer this, many of which are grounded in hard numbers and the small fractions of budgets devoted to space. But I tend to favor a far more human-centered abstraction that zeroes in on our innate and perhaps evolutionary drive to explore. When we stop reaching, we stagnate. Imaginations dim. The muscle of adventure atrophies. And our problems often grow. When the spirit of a people is mired only in fixing what's wrong at home, or worse—in building walls, be they physical or metaphorical—the ability to pursue what *could be* fades. In fact, that is precisely the cautionary tale now driving many of China's current ambitions. Senior Chinese leadership often recall the effects and exploitations suffered when their country—once the world's most technologically advanced—turned inward, and were left vulnerable in a period of Chinese history known as the century of humiliation.

Rules and systems, in fact, tend to be set by those who go out and build them, even if history itself is littered with examples of those who have wreaked havoc and exploited the very worlds they discovered. That doesn't mean we shouldn't chart and settle the unknown. It does suggest, however, that we should learn from our mistakes, accept that some will inevitably venture out, and acknowledge that it is wiser to have trusted stewards at the table than to surrender control to those who may do us harm. If we cede space, if we forgo our natural explorer instincts, if we fail to invest in the kinds of infrastructure and partnerships needed for the coming global and space economies, we may ultimately wake up to a system built by our adversaries, designed not to serve but to dominate.

If pursued, however, today's fragile and uneven space efforts could spark something transformative. Earth may even one day be analogous to those early savannas where humans first evolved. Just as modern society seems light-years apart from its early migratory ancestors, a space-born population could also diverge in ways that defy imaginations. As we fuse with our machines and probe the quantum realm, space could indeed become the crucible where scalable, high-energy, and data-rich platforms converge with just the right kind of environment needed to reshape our realities both on Earth and throughout the cosmos. Yet beyond questions of control, conflict, and falling debris lies a deep uncertainty about how those systems will evolve, whether they can coexist within the social fabric that we've come to know, and where that power will ultimately reside.

At least one thing, however, is clear: All of this is only just beginning. And that offers us a fleeting chance to set the course. Now is the time for a more engaged seat at the table, given whoever dominates space will also be poised to oversee our lives here on Earth. In this new contest, a lingering question has taken on greater urgency: Will this emerging system be one driven largely by profit and corporate interests, or one of authoritarian control that suppresses dissent? Perhaps, through deliberate action, a third path might be forged.

But the hour is getting late.

As priorities shift and a tech elite drives a consolidating industry, Beijing is buoyed by the most powerful leader China has seen in a generation, who has described space as his nation's "eternal dream," and is working to upend the U.S.-led space order. That means if Western leadership fails to respond with strategic urgency and sustained investment, or focuses merely on a strategy of flags and footprints as it did during the age of Apollo, it risks forfeiting what could be decisive influence in an emerging market that now boasts trillion-dollar valuations. But perhaps more importantly, it could also signal a shelf life for the kinds of democratic ideals upon which much of Western society is based. For now, however, space remains an open canvas of possibilities: not merely a source of wonder and strategic importance, but also a reflection of who we are, and perhaps what we may become. It's time for deliberate choices about the people, rules, and systems that we allow to win out. It is, after all, a choice.

Yet, admittedly, the path forward will not be easy.

Humans, made as we are and saddled with self-interest, aggression, and innate tendencies toward tribalism, have—at best—a checkered history when it comes to competition, expansion, and coexistence. Out there, the impacts of war, social disruption, corporate abuse, environmental degradation, and even questions of human obsolescence are magnified. But there are ways to work together, provided we collaborate from positions of strength and are backed by the force-multiplier effects of our allies. If we are to bask and safeguard ourselves from the many benefits and dangers of operating in orbit, we'll need a global framework that bares the teeth of enforcement. Perhaps the commercial sector offers a way forward, framed as a means of global collaboration to make space safer and more sustainable, so long as it does not amass the kind of unchecked power that enables it to hijack the very systems that keep authoritarians at bay.

Unlike the dinosaurs, we have been afforded a chance to steward our own longevity and choose our systems, provided we capitalize on the currency of deliberate action. But in this pivotal moment, it's important not just to understand where ambition and technology intersect, but also to identify those who are driving such endeavors.

This book therefore is meant to serve as a kind of road map, offering a rare glimpse into today's space race, which includes stories and perspectives from across five continents, including those from *both* sides of the Pacific. It is *not* a review of the usual suspects. Rather, it is an attempt to weave together the threads of a truly global story by focusing on those often quieter figures, who are nonetheless ushering in a brave new era. To fully appreciate what they're doing and where they're taking us demands also a sense of where we have been.

As such, this story commences with a nod to history, atop the very launchpad that delivered the first humans to the Moon. There, a new generation of NASA veterans are again taking lunar aim, only this time to engender a more lasting human presence beyond Earth at a moment when the human species is itself undergoing profound change.

Cape Canaveral is where we begin, with a bit of chocolate, a flute of champagne, and a man from Iran, who as a boy descended from his desert rooftop to marvel at a neighbor's small television in the summer of 1969. There, his eyes would devour the grainy black-and-white images of humanity's small steps and giant leaps.

## • PART I •

# RETURN TO THE MOON

The Falcon 9 rocket that would deliver *Odie* toward the Moon, February 15, 2024.

· CHAPTER 1 ·

# LAUNCH DAY

On the eve of America's return to the Moon, a man from Isfahan walked in from outside. He had been standing on the balcony of NASA's Operations Support Building II at Kennedy Space Center, gazing out at the faint silhouette of a Falcon 9 rocket, just as vapor trails twisted their way up the launch tower. Inside, a crowd of NASA officials mingled between an open bar and an ever-dwindling chocolate-filled buffet. But mostly, they were just waiting. A count-down clock on the wall now showed launch was only an hour away. And Dr. Kam Ghaffarian, who had co-founded the very company responsible for that lunar lander tucked inside the rocket's payload fairing, was about to offer up a few words.

Like many children of the 1960s, he had grown up in a time of social and technological upheaval, where dreams of spaceflight mixed with political turbulence at home. His path to become an American citizen and contractor to NASA had been an uncertain one. Yet now, hair white with age, Kam stood beside Intuitive Machines' co-founder Steve Altemus, myself, and a small cohort of others in the cool of that Florida night, awaiting the first steps of a plan to open the Moon for business. At the time, only three national programs had ever landed there. But never this far south. Never had a private company done it. And never had a rival nation planned to build such a lasting presence. Prestige, material gain, and influence were all at stake. But there was something else, something far more influential to the future of Earth and those who governed it. More than just the Moon, space itself had quietly reemerged as a frontier and gatekeeper to humanity's AI-driven future, offering solar energy, abundant resources, and natural cooling for the very systems upon which future generations would rely. China, which had already surged ahead in critical industries, such as batteries, solar cells, and rare earth elements, was busy with parallel strategies in robotics and quantum technologies. But space seemed to be the arena where those ambitions converged. And certain locales were key. Leadership in Beijing knew that and was targeting the same shadowed region near the Moon's south pole as that Texas company now poised for launch at Kennedy Space Center.

The stage had been set for a new kind of space race, where state-led ambitions in the East were colliding with commercial ventures in the West, albeit under the umbrella of governments in the United States and Europe. The Moon's south pole, perilous and unforgiving, had become a strategic prize that fit into those ambitions, whereby future dominance would arrive through sustainable control of the kinds of resources and networks upon which others would soon depend. And so, as the Texas company's lunar lander, *Odysseus*—affectionately known as *Odie*—geared up for launch, it did so under the weight of that broader backstory.

Tim Crain, meanwhile, was back in Houston. As the company's third co-founder and chief technology officer, he would remain in mission control to support liftoff, staging, and separation. The effort was part of NASA's broader shift from architect to client, funding space companies like Intuitive Machines, SpaceX, Firefly Aerospace, and Blue Origin to build the rockets, landers, and satellites it needed to compete. Still, at least one critical question remained unanswered: Could Moon landings really be made commercial?

If they couldn't, or if China could significantly outpace its rivals and use the Moon as it envisioned, the global power balance on Earth would surely be ripe for change.

Given that, perhaps it is not surprising that all this had become quite personal for Kam. Generations earlier, he had watched his old country fall under the grip of another kind of authoritarian government, and he now wanted to "ensure the Chinese never surpass us in space technology."

"Decisive action will demonstrate to our allies and China that the U.S. will not cede this leadership, or domain, and will rally other countries to join us," he wrote in a column for *SpaceNews*. For the moment, however, the United States still held the high ground. And it was gambling on a new generation of private space companies to maintain and expand that legacy.

But it wouldn't be easy. Jagged peaks and pockmarked terrain littered the lunar south, with both constant light and eternal darkness. Temperature swings spanned hundreds of degrees. Navigating that would test the very limits of modern technology. Failure not only would be a setback, but would deal a strategic blow to America's credibility and its new commercial tack, lending credence to those who questioned the strategic value of the Moon at all.

Kam knew this.

So did Steve.

But with Kam especially, who conveyed a persona both of shrewd businessman and wistful stargazer, not to mention witness to political change, I couldn't help but wonder where his mind had drifted as *Odie* prepared to blast off. As a boy, he had seen Iran during its White Revolution, marked not only by a consolidation of political power but also by a series of modernization efforts that included sweeping land and literacy reforms, as well as women's rights. Western music, film, and fashion swept Iranian society, just as investments in infrastructure and technology led to an increase in the demand for engineers. In those early years, he was learning a lot; the product perhaps of a curious mind and a nation in transition. And yet it was a moment in 1969 as an eleven-year-old boy that seemed to cement his otherworldly pursuits.

In fact, that night for much of the world had suddenly seemed to bring human society a bit closer to the stars.

His neighborhood—like those of many cities and towns across the globe—was abuzz with excitement. It was nearly midnight. And at night, Isfahan takes on a distinctly different feel from the daytime cacophony of street vendors and motorbikes that meander past tiled mosques and covered bridges. Especially during those hot summer evenings, when much of the world retreats indoors, many residents would escape the stifle by hoisting themselves onto rooftops to bask in the cool of the evening air. Up there, stargazers could be made.

The region has a special history with the night's sky. Early Babylonian astronomers in Mesopotamia compiled the first star catalogs, while earlier Sumerians observed the movements of the planets and recorded the names of constellations. Ancient Egyptians used the stars for alignment of their temples and pyramids, and developed calendars based on lunar and solar cycles. The desert sky, with its low humidity and sparse cloud cover, tends to make celestial objects appear more visible. And by its location in the foothills of the Zagros Mountains, Isfahan's comparatively higher altitude allows starlight to pass through fewer filters before reaching the observer's eye. In the generations before modern light pollution washed out "the floor of heaven . . . inlaid with patines of bright gold," as William Shakespeare once described it, rooftops were thought to be a choice spot for amateur astronomers.

"I was mesmerized by the stars," Kam explained. And during our conversation he wondered aloud if humanity could actually "go there."

That night, however, in the summer of 1969, Kam wasn't sprawled out across his rooftop. Instead, he had edged his way down to just outside his neighbor's house, where he peered through dusty windowpanes to witness a first for humankind. There, broadcasting on a small television, were images of Neil Armstrong's initial steps on the Moon.

"My eyes were *this* big," he told me. "That's what I wanted to do."

And yet shortly after he had begun college in the United States, his country would enter the tumult of revolution, eventually culminating in the overthrow of the Pahlavi dynasty. In its aftermath, bearded revolutionary guards roamed the streets, enforcing a new constitution and religious government. Kam, who had left two years earlier on a student visa, would watch from Catholic University in Washington, D.C., as his country transformed. Gone for good, he became an American citizen and earned a computer systems job at Lockheed, or, as he saw it, a prime chance to work with NASA. Then, in 1994, he took a risk, borrowing $250,000 against the value of his home to start his own engineering services firm as an adviser to the agency.

"It was an all-in gamble," he explained. "If I was not successful with that business, I wouldn't have a home."

Ultimately, it seemed to work. And Stinger Ghaffarian Technologies would prove lucrative, offering IT, engineering, and mission operations to NASA—and thus transforming the boy from Isfahan into one of the world's richest men. Three decades later, he had amassed not only a billionaire's fortune but also a portfolio of moon shots devoted to lunar landers, commercial space stations, modular nuclear reactors, space suits, and even a research lab dedicated to the physics of interstellar travel. It was big-picture stuff. But unlike Elon Musk and Jeff Bezos, who had each built space behemoths of their own, most of Kam's efforts tended to focus on what those humans and machines could do once *in* space. He also seemed to relish a far lower profile than Musk or Bezos; albeit with an outlook that was no less grand. And on that February night in 2024, Kam sought to lay out the broader vision in a speech not made public until this very book.

Clad in a blue blazer and coral shirt, he approached the lectern. "I want us to go forward in time within the next thirty, forty years," he began. "Imagine that you would have hourly launches to a space city, outside Earth. Imagine that we will have daily launches to the Moon where we have a habitat, and that it's a platform for future human space exploration, not only within our solar system, but outside to interstellar space within our galaxy," he described.

Today, the Moon was the goal. But with a few upgrades, there was little reason the lander inside that payload fairing could not be modified for the harsh realities of the Red Planet.

Kam knew that.

"Imagine that we have weekly trips to Mars, where we also have many people living, [where] we have a colony, and people actually live and work there," he continued. "Imagine that you also have launches, maybe monthly or quarterly, for interstellar travel outside our solar system to other places in our galaxy. . . . My personal belief is that the destiny for human beings, for humankind, is to go to other stars and find new homes in those stars. And the new chapter of that journey starts tonight."

Much was now riding on this moment, not just for the company that Kam, Steve, and Tim had built, but also for the nation at large. U.S. preeminence was no longer preordained. Intuitive Machines seemed to be at the tip of an American spear into space. And these men represented the company's public face. But there was a bigger team—many now back in Houston—who also were collectively holding their breaths.

To understand their story, and to truly grasp the moment at hand, one must rewind the clocks three years earlier to a sunbaked tarmac in Texas, where a team of engineers were testing the very propulsion system they hoped would bring *Odie* to the Moon.

· CHAPTER 2 ·

# BUILDING LANDERS

There was a problem in Houston. Deep inside the engine of what could be America's first lunar lander in more than fifty years, liquid methane and oxygen were now mixing on the edge of ignition. Once sparked, the cocktail would unleash a torrent of energy that would propel the craft through space and toward the Moon. But landing would be just the beginning. If they could do it, they'd also need to build the very systems that future missions—and future astronauts—would rely on to survive and operate. To do that, they'd have to mine the Moon for water, iron, titanium, and a curious isotope with tantalizing potential.

Sown into the Moon's surface by solar wind, helium-3 is thought to be a near-ideal fuel source for fusion reactors, though its more immediate value lay in its ability to absorb heat from quantum computers to protect the delicate information they stored. Extracting the isotope, processing it, and transporting from Moon to market, however, left many skeptical; a business case yet to be proven.

But that wasn't the problem. Not today at least.

First, *Odie* would have to get there. It was 2022—two full years before that moon shot from Cape Canaveral—and this Texas crew of engineers now faced a critical issue with the craft's propulsion system. Propellant was now funneling through the metal channels of a heat-defying alloy known as columbium, where—upon ignition—temperatures half as hot as the surface of the Sun cooked inside. Cryogenic cooling would keep the engine itself from vaporizing, just as that powerful blue flame roared to life, bursting out the back of the craft. To the untrained eye, it all looked good. The vehicle thundered and rattled. Yet the metal composite was holding, a triumph of the 3D-printed engine that was now working as intended.

In truth, however, it wasn't.

And Rob Morehead knew it.

Having served as engineering lead on a similar NASA prototype, known as Project Morpheus, he could sense something was off. A decade earlier, Rob had watched Morpheus erupt into a ball of flames during a test flight at Kennedy Space Center, thus fortifying the arguments of the old hands at the agency who still preferred those presumably safer, tried-and-true hypergolic propellants. And for good reason. They were easier to work with. The mix ignited on contact. And the fuel could be stored at room temperature.

So, why mess with that?

There were a few reasons: cost, toxicity, complexity to produce, and perhaps above all efficiency. In commercial space, where cargo is king, smaller, more efficient engines could be the difference between a profitable craft and one destined for the back of the line. Of course, none of that mattered if the engine didn't ignite. And in the near vacuum of space, methane was still largely untested.

If it didn't, *Odie*'s mission would be over before it began.

Still, the risks seemed surmountable. And so Rob would turn in his old NASA badge to join the private sector, where presumably he'd be given more leeway. Yet questions of time and money also now took center stage. Unlike the old days, where cost-plus contracts could effectively incentivize bloated budgets and yawning delays, this new era of fixed-cost contracts meant company margins depended on the ability to squeeze out more with less. The idea was to move quickly, reduce costs, and nudge into existence the beginnings of a lunar services market, even if many privately wondered whether such a market—devoid of taxpayer subsidies—could ever really exist on its own. Then again, now was a scrappier time. Barriers to entry were dropping. Engineers no longer needed fortunes to put their machines in space. Plus, in business, availability often mattered just as much as ambition.

And *that* was the other reason for methane.

Not only was it found naturally on Mars, but it also could be created there, using a more than century-old process invented by a French chemist who discovered that when hydrogen was added to carbon dioxide under certain conditions, it produced methane, water, and energy. Given that Mars's atmosphere is filled with carbon dioxide, and with water ice available in the soil, methane fuel stations could be built on location, thus reducing the amount of return fuel a ship would need to carry from Earth. In that sense, the Red Planet—not the Moon—seemed to be the real goal. For Rob, and many others across the industry (including those at SpaceX and in Beijing), that meant methane was no longer just optional. It was a key ingredient in this new era of space travel. Still, Rob might have just stayed at the agency, had it not been for an impromptu meeting over beers between a lizard and a pirate.

It was Halloween, circa 2009.

He was dressed as the Geico Gecko. The swashbuckler was Steve Altemus, who—before co-founding Intuitive Machines—had overseen human spaceflight as Johnson Space Center's deputy director. The two men had been friends and colleagues. And they knew each other from those old Morpheus days. But what Steve didn't know was that his former engineering lead had been not-so-quietly teaching high school students to build rocket engines, with parts large enough to power the Morpheus lander in his garage. He had figured that if NASA wouldn't go in on methane, he would just try it himself.

Steve's eyes widened.

"Show me," he said, as the two men pored over the particulars. The core players of their company now seemed to be coming together.

Years earlier, during a 2012 dinner in Washington, D.C., Steve had met with Kam to lay out the vision. Back then, he had suggested devising a company to build drones and flight systems and to use their own NASA intuition to meet emerging technology demands in the private sector—hence the name Intuitive Machines. "It would be a shame for the music in your head not to be played," Kam replied, as the three men forged the beginnings of what would become that Houston-based company. Rob would later join them, even as Steve cautioned that "this wasn't NASA."

But maybe that was the point.

In fact, Tim and Steve had begun considering a private company while still at the agency, when human spaceflight "was not the top priority," Tim later explained, having observed yet another political shift and change in priorities with what was then the incoming Obama administration.

An independent review carried out by a group known as the Augustine Committee reinforced that thinking as early as the fall of 2009, finding NASA's crewed spaceflight program, Constellation, grossly over budget and years behind schedule. The report called it "unsustainable" with "insufficient funds" while seeming to suggest a private-sector-based approach, which, it said, at least "has the potential to stimulate a competitive commercial space industry." Constellation was effectively dead, with a pledge to extend the life of the International Space Station (ISS) and commitment of $6 billion over five years to support private companies. In this way, despite their differences, both the Obama and later the Trump administrations seemed to converge on a similar approach, leaning more heavily on the private sector while phasing out legacy programs.

The pushback in both cases was fierce.

During the Obama years, Apollo luminaries Neil Armstrong, James Lovell, and Eugene Cernan penned an open letter that admonished the administration for "grounding JFK's space legacy," and called on Congress to "overrule this administration's pledge to mediocrity." Those like Senator Bill Nelson (D-Fla.) and Senator Richard Shelby (R-Ala.), who both held long space industry ties, bristled at the notion of a new client model that opened the door to then upstarts like SpaceX, whom Shelby referred to as mere "rocket hobbyists."

Charles Miller, who would later serve as part of the Trump transition team for NASA, described that pushback as part of an "iron triangle," consisting of NASA bureaucrats, big aerospace companies, and politicians on Capitol Hill, which "locked [the United States] into a 1960s socialist paradigm." Those like Miller saw a more open market approach as a principal means of breaking free from industry stagnation. Now, those voices were getting their shot. The industry seemed poised for growth, with a bevy of new competitors adding to trillion-dollar market projections.

In 2017, Vice President Mike Pence called the Moon "a stepping-stone" and "a venue to strengthen our commercial and international partnerships," effectively cementing the emerging Artemis program and America's return to the Moon as a pathway to Mars.

For Steve, that was all he needed to hear.

"We completely reoriented our business model to focus on the Moon," he told me, with methane-based propulsion systems that could one day be used for the Red Planet. Two years later in March, President Trump made it official, formally announcing that America again had plans to return to the lunar surface, only "this time, we will not only plant our flag and leave our footprint, we will establish a foundation for an eventual mission to Mars and perhaps, someday, to many worlds beyond."

Now the private sector had to deliver.

· CHAPTER 3 ·

# THE VIEW FROM CHINA

A world away, a man in Beijing named Xue Suijian had tuned in to watch *Odie*'s launch from Cape Canaveral. The former deputy director general of China's National Astronomical Observatories was an astronomer and astrophysicist by training, usually more focused on the observing of distant galaxies. Curiously, however, he had a stake in *Odie*'s success. Despite growing rancor between the United States and China, his old observatories were part of a research institute operated by the Chinese Academy of Sciences (CAS) and part of a collaboration between the two nations that seemed to be grandfathered in from a far different time. Make no mistake, CAS was still at the forefront of China's broader space efforts, including the Tianwen Mars explorer missions and its burgeoning Chang'e lunar program, doing the things needed to expand China's growing lunar footprint. In fact, when it came to the Moon, CAS would publish the highest-resolution geological map of the lunar surface to date, which included more than twelve thousand impact craters and eighty-one impact basins, useful for future mining and development of a ground station. The first phase of that proposed facility—a joint project with Moscow—had been scheduled for completion by 2035. But a more "comprehensive lunar station network" included "exploration nodes on the lunar equator and the far side of the Moon," explained Wu Yanhua, chief designer of the Chinese deep space exploration project. Wu wanted station components capable of operating across a sixty-two-mile radius, given the effects of lunar landers. Engine blasts inevitably created sprays of abrasive, electrostatically charged regolith that can damage surrounding structures, craft, and space suits. Such an expansive radius, however, had another ancillary benefit, effectively laying de facto claim to strategic lunar terrain over valuable mining sites and ice-rich craters. Having already retrieved helium-3 from the Moon's far side, China now had bolder ambitions, which included not just a sustained human presence on the Moon, but a far more comprehensive system of networks and supply chains for much of the near solar system.

By 2021, the nation had made major strides with its Tiangong space station, achieving breakthroughs in artificial photosynthesis to support long-term life in space, while pursuing space planes, mining operations, and access to unique points of interaction with gravity—known as LaGrande points—which allow spacecrafts to remain relatively stable with minimal fuel use. Virtually all of it aimed at securing resources, pursuing science, and establishing strategic dominance. Central to that vision, however, were at least two key elements: data and energy, with plans for a kilometer-wide solar array already in the works. These were the grand, Hoover Dam–like ideas, the kind of things for which America used to be known. And there was no telling just how much power and influence they might afford.

But there *were* hints.

Long Lehao, a leading Chinese rocket scientist and chief designer of China's Long March rocket program, claimed that such a massive orbital solar project could be capable of matching the total energy derived off Earth's annual oil output, which would then also complement burgeoning AI-driven orbital supercomputers that his country was fast developing. Many in the United States were skeptical. But China's long-term visions were increasingly clear: energy, AI, resource extraction, research, and astronomy, a launchpad for deeper space endeavors, and finally a strategic network linked to its growing operations in Earth's orbits that many in Beijing hoped would press a clear Chinese advantage. To keep that advantage, China was both shoring up global supply chains and critical raw materials, as well as future talent pipelines. By 2024, the nation had for the first time outpaced American efforts when it came to venture capital for space technology start-ups, while also nearly doubling the annual number of graduating STEM PhDs relative to the United States. Meanwhile, China's military space bureaucracy was also fast evolving, having undergone several major reforms, with three reconstituted forces known as the Cyberspace Force, the Information Support Force, and the Aerospace Force, the latter reminiscent of America's newly minted Space Force. With newfound infrastructure and resources in orbit, there was a growing need to aggressively defend those interests. Chinese satellites, according to U.S. officials, were now performing coordinated "dogfighting" maneuvers in low Earth orbit. Space was very clearly a war fighter's domain.

Still, for all its expansions, China's space program was still young. By 2025, its large, reusable rockets were still in their nascent stages, just as the global launch sector remained overwhelmingly dominated by the U.S.-based SpaceX. By then, SpaceX rockets were on pace to be launching every third day, and accounted for well over half of the total global launch cadence, giving American leadership—at least for the moment—an unparalleled resource when it came to getting their machines in orbit. That had manifested in the world's most comprehensive satellite constellation, albeit with a disproportionate reliance on just one company, whose CEO was growing more powerful across multiple fronts.

China's eyes on space were accordingly trained on Musk.

"On WeChat . . . every Starship launch, every Starlink launch, and frankly every word Elon says is covered with interest," explained Blaine Curcio, founder of Orbital Gateway Consulting. "If you line up all the big milestones that SpaceX has had, and what China is doing, there's a bit of a lag, but there's a clear mimicking." Still, that imitation didn't preclude moments of overlap. On occasion, there was even room for collaboration. Despite the rhetoric, the hawks in Beijing and Washington weren't the only voices with influence.

Xue, in fact, was a prime example.

It took me months to reach him. But when I finally did, it came in the form of a middle-of-the-night phone call from Beijing.

I had been expecting it.

Still, when that sharp, sudden ring pierced the dark of night, it felt like a pit had simultaneously formed in my stomach. There was a pause and the crackle of a faraway line when I answered. Then, in a measured voice, he spoke.

"David, it's good to meet you."

For a time, Xue had held one of the most influential positions in Chinese space science, having championed major international collaborations and played a key leadership role in turning his nation's celestial ambitions into tangible realities. Soft-spoken and reflective, he represented the beginnings of what would become an insider's look into China's space efforts. And I knew I couldn't waste it.

At first, the conversation hewed along expected lines, consistent with the Chinese Communist Party. There was a game of sorts to play here, and perhaps even a mutual awareness that we may not be the only ones listening in on that call. The former deputy director general, likely aware of that dynamic, couched the conversation in patriotic terms, while calling America's commercial pivot a "clever," though smaller scale tack. "You can hardly make the comparison with China's lunar missions."

"China is a national-scale invested effort," he emphasized. President Xi had also effectively deemed space-based systems critical infrastructure, all but ensuring their continued funding, unlike the more fickle nature of American space policy.

Yet as the hours pressed on, our conversation took an unexpected turn. More than a decade earlier, Xue explained, he had helped broker one of the first collaborations between the U.S. and China, having inked an agreement with a Hawaii-based observatory that would prove relevant in this modern race to the Moon. The rivalry, in fact, hadn't always been so contentious.

At the turn of the twenty-first century, China and the United States had grown economically interdependent, despite concerns over trade imbalances, intellectual property rights, and currency manipulation. In spite of recurring issues, like human rights abuses against the Uyghurs and other ethnic and religious minorities in China's Xinjiang province, as well as territorial disputes over Taiwan and the South China Sea, cooperative efforts to address other major issues like climate change and terrorism had presented real avenues for collaboration. Relations would, of course, later devolve into a series of trade wars, escalating cyberattacks, espionage, and ever more hawkish commentary. Although, curiously, none of that seemed to undermine Xue's hope that data gleaned from the IM-1 launch would return an unlikely courtesy that he and China's lunar program had extended to American astronomers more than a decade prior.

The year was 2013, a seminal one for China's young space program.

Beijing was about to try its first-ever Moon landing. After a tense thirteen-day journey and insertion into hundred-kilometer lunar orbit, its Chang'e 3 lander then proceeded to etch its name into the annals of history, joining the exclusive club of Moon-faring nations. The last soft landing on the lunar surface had, in fact, been carried out by the Soviet Union in 1976, which is, coincidentally, the same year that Mao Zedong had died.

Now, a new chapter both in China and for space was unfolding.

After touching down on the lunar plain known as Bay of Rainbows, the craft released a rover called Yutu (or Jade Rabbit), which—in Chinese folklore—is a companion to the Moon goddess Chang'e. But that wasn't its only payload. For astronomers in Hawaii, the bigger interest on board was an ultraviolet telescope. Intended for galactic observations, benefiting from far less radio disturbance on the lunar surface—which allowed for less obstructed studies of neutrinos, cosmic rays, and distant planets—the device would soon capture images of the pinwheel swirl of the center of the Milky Way galaxy, some twenty-one million light-years away from Earth. Tantamount to a temporary Chinese observatory on the Moon, those images were then shared with American astronomers in Hawaii.

A decade later, with *Odie*'s launch, those same Hawaii astronomers sought to return the favor by allowing their Chinese counterparts, unbeknownst to the IM team in Houston, to use data they expected to gather from two cameras they had secured as payloads aboard

*Odie*. Steve Durst, head of that Hawaii team, had helped spearhead that effort, working closely with Xue.

"He convinced me it's a good opportunity to collaborate with each other," Xue explained of Steve, who had routinely traveled to Beijing to meet with his Chinese counterparts. "It's pretty rare to have that kind of collaboration."

The politics of the moment, however, seemed to have other plans.

· CHAPTER 4 ·

# PROBLEMS IN HOUSTON

Even without the sear of rocket tests, that day in Houston would still have felt especially hot. Heat waves radiated off the asphalt near Ellington Air Force Base, where Intuitive Machines had by 2022 only recently opened its factory. Thirty-five hundred pounds of force scorched the range. But it was unavoidable. *Odie*'s propulsion system needed fixing.

A "flame range" facility later opened in 2023 consisting of a thirty-eight-hundred-square-foot reinforced-concrete chamber surrounded by a twenty-five-foot-high perimeter wall, where Intuitive Machines' engines are now tested. In 2022, however, it was effectively out in the open. Back then, I traveled to Houston, Texas, to meet Steve Altemus, Tim Crain, and Rob Morehead at the Intuitive Machines factory, the first of several meetings over the years.

As Steve approached, the team worked the problem from behind a concrete blast wall. Rob, meanwhile, listened on the headsets.

"How's it going?" Steve asked.

No response.

"Rob?"

He seemed lost in the moment, having noticed an anomaly. Multiple cameras were broadcasting a live feed of the rocket's exhaust. Rob was leaning over a series of screens that flashed readouts of the engine's performance; his hands by now pressing those headsets more firmly against his ears.

"We call him the rocket whisperer," Steve said, a moniker earned for his uncanny ability to detect even the slightest irregularities in the engine's rumbles and sputters. A matrix of numbers flashed across four large monitors. Rob looked stressed. Of course, so did Steve. Affable with a kind of back-patting demeanor, he was nonetheless under pressure. His company was not the only one NASA had enlisted. Astrobotic, Firefly Aerospace, and others were aiming for the distinction of being the first company on the Moon. Failure could render them obsolete. He needed to know what the problems were, especially today. Around the corner at company headquarters, a smattering of political leaders, national guardsmen, and entrepreneurs had gathered for an event to discuss Houston's future. The ideas were big: visions of traveling to Tokyo in under an hour, which once again seemed to focus on China. A Beijing-based company called Lingkong Tianxing Technology was already in testing of a commercial supersonic passenger plane expected to travel at a spellbindingly fast Mach 4, or roughly twice the speed of the Concorde, just as China's broader space program was gaining ground.

Everyone there seemed to want a status update. With market share and industry reputation on the line, a mix of excitement and stress was building. It seemed impossible, however, not to recognize where others had stumbled.

For many companies, the title of first commercial Moon landing had been tantalizingly close.

In 2023, a software malfunction was all that had prevented a

Tokyo-based company from becoming the first, reportedly causing the spacecraft to plummet at a staggering rate of more than a hundred meters per second before it smashed into the Moon. That crash followed an earlier Israeli attempt, which had been carrying what might have been the Moon's first residents: thousands of near-microscopic aquatic animals called tardigrades that can survive in all kinds of unlikely places, including (at least for a while) the vacuum of space. Sadly, the vehicle's gyroscopes failed—among other things. And the tardigrades were presumed dead.

"We knew there were risks," explained Nova Spivack, founder of the Arch Mission Foundation behind that Beresheet Israeli mission. "But we didn't think the risks were that significant."

History, however, tells a very different story. Despite the six crewed U.S. landings between 1969 and 1972, most attempts ended in failure, including NASA's first six attempts to land a robotic lander on the Moon. The Soviet's Luna 15 mission—a last-ditch effort to return lunar samples before the United States—even crashed onto the lunar surface while Neil Armstrong was standing on it, just a few hundred miles away. Generations later, both India and Russia had also crashed their respective landers near the south pole before the Indian space agency finally touched down its government-run *Vikram* lander in August of 2023.

Still, no commercial craft had ever done it. The more "shots on goal," as Thomas Zurbuchen, then NASA's associate administrator for science, put it, the better. "A reasonable expectation is a 50 percent success rate at the beginning."

"We're not trying to redo Apollo," Joel Kearns, the deputy associate administrator for exploration in NASA's science mission directorate, later explained. Still, before any of it might happen, *Odie*'s engine had to be sorted.

"Rob?" Steve repeated. "How's it going?"

"It's going," he replied finally. It was the landing sequence. The propulsion system was meant to employ a constant burn for a stable lunar descent, avoiding the pulsing technique that might have destabilized attempts by other landers. Unlike setting down an airplane on Earth, which employs aerodynamics and lift to coast to the sur-

face while under the influence of gravity, landing a rocket on the Moon involves using its engines essentially as brakes and powering down to roughly one-fifth of max thrust. That, of course, also runs the risk of stalling. In the past, to solve this, craft employed blips of power. But turning on and off the engine can also create a sloshing movement of the fuel, and throw off the craft's equilibrium. Plus, *Odie* would be fully automated at that point, and there were so many things that could go wrong. Just one of them could compro-

mise the mission: A loss of fuel flow, imbalances between the fuel and the oxidizer, extreme thermal fluctuations could all result in a crash, or a drift off into space. And at the moment, the team wasn't getting the right mix of thrust and stability it needed.

If *Odie* were to survive the Moon, the path would have to go through Texas.

· CHAPTER 5 ·

# A NEW SILK ROAD

Some five thousand miles south of Houston, Xue stepped off a plane and in to the Argentine capital. There, he had entered a decidedly different world, far removed from the cowboy-hat-wearing engineers in Texas, and even farther from his home in Beijing. Buenos Aires exudes a kind of romantic nostalgia, strewn as it is with cafés and bookstores, French balconies, and an Italian lilt that often slips through a Spanish tongue. The Chinese capital, by contrast, boasts a more vast, imperial demeanor, with wide boulevards and ancient hutongs, leaning on a history that spans thousands of years and hints at a more dynastic past.

At first glance, the two locales could scarcely be more different. Yet there was mutual interest here, especially when it came to space. Xue knew that. In fact, he had been dispatched on a diplomatic mission at the behest of senior Chinese leadership, tasked with helping to open this Andean nation to China's broader space endeavors. To build a global network, Beijing needed international partners, not just for tracking and telemetry, but also for a growing body of astronomical science—much of which Xue was now spearheading. That would require partnerships and buy-in. But there were other, more practical necessities. And when it came to China's space interest in Argentina, the focus now centered on building what would arguably become among the most prominent symbols of Beijing's growing influence and power in the Western Hemisphere.

"At that time, China didn't have any giant radio dishes to receive the data from the Moon," Xue told me one evening from Beijing.

The Southern Hemisphere was crucial in that respect, because it offered novel views of the night sky, particularly in the more rural south, where the Andromeda galaxy burns so brilliantly that it can make an amateur astronomer out of just about anyone. There, the galactic center and the Magellanic Clouds, which consist of a pair of dwarf galaxy companions to the Milky Way, are visible even with the naked eye. But beyond astronomy, it was also strategically significant. To maintain connectivity with spacecraft and satellites, China needed a massive deep space ground station to help coordinate its efforts. Without a direct path from a transmitter to a receiver, communications are delayed or even cut off at different intervals of the day. Ground stations bridge critical gaps in line-of-sight coverage over the South Pole and midlatitudes, at least until orbital infrastructure could reduce those dependencies on Earth-based systems. And to that end, by the mid-2010s, top Chinese military and space officials had begun making frequent visits to South America. Xue's deployment was indeed part of a multipronged effort that had started decades earlier.

Building on a 2004 state visit to Buenos Aires, during which China's president, Hu Jintao, and Argentina's president, Néstor Kirchner, signed a series of bilateral agreements—including the framework for cooperation on the peaceful use of space technology—the early stages of collaboration were now taking shape. Argentina was an attractive partner: a relatively stable country in need of cash, and home to useful resources, especially lithium, which could be employed in the kinds of high-energy-density batteries used in spacecraft and satellites.

But at first, Beijing's approach seemed like it might cause some problems. China's People's Liberation Army (PLA) oversaw most of its space activities, including much of its nascent commercial industry, which meant it was not just those like Xue who were to oversee and negotiate. It was also military brass.

"It is all operated by a PLA unit," Xue told me one evening, regarding the development and maintenance of China's new ground station in Argentina, a hulking $50 million facility that sat atop a two-hundred-hectare site and was employed in coordinating a seminal Chinese acheivement in space: humanity's first-ever landing on the far side of the Moon.

"Actually," Xue said, almost reflectively, seeming to ponder a different system that many in the country's space leadership had quietly suggested, "I would like to suggest China to learn from the system of the U.S. to try to separate" civilian from military activities in space.

"For example," he added, "NASA from the start was set up as an independent space administration to conduct purely civilian programs. If we did it that way, it could be conducted with equal collaboration with the United States. Otherwise, it's very difficult. Every national-scale space program is managed by a national security department."

Xue was effectively calling for more collaboration. Still, the optics of engaging directly with the PLA would remain a roadblock, especially for the United States, where U.S. law restricts NASA's cooperation with China. Joint efforts between civilian scientists were one thing, provided exceptions could be made or grandfathered in. Cooperating with the Chinese military was an entirely different matter. And yet it wasn't just Chinese officers, scientists, and engineers descending on this suddenly space popular nation. There were others who seemed just as determined. Former Soviets had begun appearing, too, their knowledge hard-won and still highly valued. In fact, long before ever meeting Xue, I had dinner with one. But before we could be introduced, I was instructed to bring a gift.

Stanislav Makarchuk liked vodka, the imported stuff.

And before that Friday afternoon meeting at Gran Taberna, a checkered-tablecloth locale in the heart of Buenos Aires, I didn't know much more than that. Only that he worked for CONAE—the country's civilian space agency—preferred Stoli, and was someone I "had to meet." Born in Moldova, a landlocked former Soviet republic of rolling pastures, monasteries, and imposing concrete that holds the distinction of being Europe's least visited country, Stanislav was ready to leave by his teenage years, studying engineering in St. Petersburg before eventually working for Roscosmos, the main successor to the Soviet Union's old space program, hobbled though it was by corruption and mismanagement. What happened next is a bit murky. But Moscow needed partners. And Argentina's complicated relationship with the United States, and newfound importance in this burgeoning second space race, made it an attractive choice. The two countries in 2019 signed a space cooperation agreement, which included aspects of navigation, research and development of space equipment and piloted spacecraft, and the legal basis for future joint space projects.

"We are interested in developing navigation projects in Latin America to have good global coverage of all signals," explained Roscosmos's deputy director general Mikhail Khailov in a statement.

But Argentina was also vulnerable. The country was on the verge of yet another default, with inflation approaching 40 percent and rising, effectively undermining what stores could keep on the shelves, like, for instance, imported vodka. When high-end products showed up, it was a good idea to buy in bulk, lest it cost twice as much in a few days. The peso might be Argentina's currency, but it was the more stable greenback that had real value.

And yet that was also changing.

After a series of controversial International Monetary Fund (IMF) bailouts, Argentine leadership had grown skeptical of Western lenders, just as bilateral trade with China had swelled from $2.3 billion in 2001 to $26 billion in 2022, with more than half of Chinese commercial bank loans in Latin America having gone to Argentina. Beijing and Buenos Aires were clearly getting closer, furthered by senior leadership at CONAE, which now included Stanislav, who by then had traveled to China three times on diplomatic missions.

"The Chinese work with us," he explained to me over dinner. "The Americans compete with us."

Referencing a host of Chinese infrastructure, he had homed in specifically on that Chinese-military-run ground station in Patagonia, which was now actively being used by Chinese space authorities, and to some extent Argentine ones.

"What we managed to accomplish [in building this station] changed the math," Stanislav explained. "It opened the door" to both Argentina's strategic significance in space and China's broader ambitions. Xue and Stanislav, in fact, ran in similar circles. And during his three visits to China, Stanislav had met directly with Xue's colleague Ran Chengqi, who would later serve as a key interlocutor in the development of BeiDou tracking stations—part of a growing Chinese rivalry to the U.S.-owned GPS navigation and timing system. Often referred to as China's Space Silk Road, or the BRU Space Information Corridor, the initiative sought to employ satellite-based navigation systems and 5G and 6G networks to connect infrastructure across Europe, Africa, and Latin America to provide for global positioning, navigation, and tracking. Naturally, as a result, concerns in Washington were growing. A 2023 House Foreign Affairs Committee analysis posited that by "embedding and serving as the Global Navigation Satellite System provider globally, [China could] influence third countries and control shares in the satellite navigation industry and service markets upon which third countries rely."

Normally, those kinds of red flags might have stalled cooperation. But Buenos Aires needed the cash and was obliged to entertain not just space initiatives but also a host of other infrastructure projects across the country. China's Zijin Mining Group Co. was already in talks to develop the nation's massive lithium reserves, despite growing protests from local opposition concerned about a lack of control across Argentine territory. Already a major global player in copper and gold, the mining giant was both state aligned and strategic in China's global expansion, which seemed part of a broader effort to shore up critical supply chains, having offered infrastructure investments and contracts that were hard for cash-strapped governments to refuse.

To some, it was a lifeline.

To others, it was déjà vu.

In other parts of the world, including within China itself, the consequences of unchecked resource extraction had already played out with devastating effects. Inner Mongolia, a semiautonomous region within northern China, for instance, was home to the largest deposit of rare earth resources in the world. Yet resource extraction there had spawned toxic lakes, polluted skies, and was reportedly responsible for the lingering smell of sulfur in places like Baotou, the region's largest industrial city, with a kind of dystopian undercurrent described by the journalists who visited. "It feels like hell on Earth," as BBC journalist Tim Maughan wrote after visiting the Baogang Steel and Rare Earth complex. The Bayan Obo mines to the north are, in fact, believed to contain some 70 percent of the world's rare earth deposits, which funnel their way onto the global market and power everything from smartphones to batteries. But all of that would quietly come at a steep environmental and human cost. In fact, the devastation was so severe in some parts that certain regions had been unofficially designated as "sacrifice zones," or places where the ecological destruction and human suffering were just generally accepted as the price of progress for global technology.

That was the concern in Argentina, which accounts for an estimated one-fifth of the world's lithium reserves. Lithium is used in batteries, thermal management systems, and even hybrid propellants within the space sector, and with the global energy transition under way there were expectations of a forty-fold increase in demand by 2040. The so-called Lithium Triangle within the borders of Argentina, Bolivia, and Chile in the salt flats of the Atacama Desert and neighboring regions had attracted international mining efforts as demands for cobalt, nickel, and manganese swelled. Local residents openly worried that their homes might be destroyed by profiteers in search of the lucrative metals. In one instance, Argentine workers even surrounded a Chinese mining company's compound and set tires ablaze to blockade the entrance.

*That* seemed the backstory here.

Argentina was rich in resources and uniquely positioned for space. But it also faced growing financial pressures that made it susceptible to deep-pocketed international partners. Stanislav, for his part, was well aware of the balancing act at hand. From his old days in Russia, he had worked for one of the few space agencies that had made real accomplishments in the cosmos. In that context, turning to a former Soviet might have just made logical sense, given it was the U.S.S.R.'s storied legacy that offered both credibility and capability at a time when Argentina needed both. It was the Soviets who had

A Chinese-military-run deep space ground station, encased by an eight-foot-tall barbed-wire fence, amid the Patagonian landscape, outside the Argentine town of Las Lajas. Visits are by appointment only, and foreigners rarely gain access. And yet a carve-out in the operating agreement allows Argentine officials access to the station 10 percent of the time, which—with the right connections—affords insights into its inner workings, 2021.

put the first satellite into orbit, launched the first lunar probe, and placed the first human in space—not to mention were the first to land on another planet.

The aerospace manufacturer and launch service provider Rocket Lab had more recently set its sights on Venus, with the goal of probing its atmosphere for organic molecules. But it was the Soviets who had delivered the first unmanned spacecraft, Venera 7, to the Venusian surface decades earlier. In a conversation I later had with Rocket Lab's CEO, Peter Beck, Beck described those early Soviet missions as "crazy" and "a massive scientific accomplishment."

"When you get lower in the atmosphere in Venus, it's almost a liquid," he explained. "So it can take hours for the probe to descend through these heavy, dense gases to land on the surface . . . you don't even need a parachute."

The Soviets had done it decades earlier. In fact, back then, even the first basic space station was forged by the connecting of two Soyuz vehicles. Now, however, with Roscosmos in various states of disarray and underfunding, the main competition to U.S. efforts had steadily shifted east, with those at CONAE helping to bring a Chinese ground station to their country's remote Patagonian steppe.

The Argentinians would provide for its infrastructure and initial services for operation and maintenance. But it was China's Satellite Navigation Office that would do the remote sensing as relations between the two drew closer. Still, the very notion of a partnership was a bit different in China, whose leadership maintained an almost philosophical bent toward self-sufficiency. Beijing didn't quite build allies in the same way that they were cultivated in the West. Rather, it developed associates. Among them had been Russia, and now increasingly Argentina.

Still, domestic politics tends to have its own way of intervening.

Just as China appeared to be doubling down on its billion-dollar investments across Latin America, something curious happened: A libertarian economist who had run a campaign on a decidedly anti-China platform would go on to win the Argentine presidency. In a reflection of growing popular sentiment, Javier Milei said that Argentina could no longer work with "communist" systems, and reportedly described the government in Beijing as an "assassin" against its people.

From a surface level, at least, a fissure seemed to be in the making. And yet pressure can also come in different ways. In fact, only shortly after his election, Milei suddenly appeared to have a change of mind, or at least a softening of the rhetoric. In reality, Argentina's economy had become hugely reliant on Beijing, with China having committed to four Argentine lithium projects, meant to invest \$2.7 billion in the country. It was a leading importer of Argentine beef and soybeans, and had been engaged in a series of major hydroelectric and nuclear projects. More importantly, China had been party to billions in financial credit provisions that had helped Buenos Aires from falling into default. After all that, it was probably difficult to just walk away.

And it showed.

After Milei's election, President Xi Jinping sent him a letter congratulating him on his victory. Milei then responded in kind. "I thank President Xi Jinping for the congratulations and good wishes," he expressed on his X account. Later, during a speech at the World Economic Forum in Davos, Switzerland, Milei seemed to have more fully changed his tune. Argentina, he warned, was in danger because Western leaders had abandoned "the values of the West."

By June 2024, he was planning an official visit to Beijing.

"He wouldn't travel to China if there was another alternative," Patricio Giusto of the Sino-Argentine Observatory told the *South China Morning Post*. "I think if he ends up travelling it's because there is no other option. If he had one, he would return all the [currency] swaps, but Milei knows he can't do that."

From Beijing's vantage, the threat had been largely contained.

· CHAPTER 6 ·

# THE PATH TO INTERPLANETARY SPACE

The road to Neuquén snakes its way through the sweeping Patagonian steppe, tracing a 150-mile path across a changing landscape toward the quiet town of Las Lajas, which—surprisingly—has a certain knack for weathering the kind of political controversy a Chinese-military-run station might stir. Many still recall a moment in 1988, following a much darker period in the country's history, when army rebels seized a nearby garrison in the hopes of igniting a national uprising.

History has not remembered this time kindly. It was, in fact, one that many in Argentina would not soon forget, but also part of the rationale for so openly engaging with America's chief rival. To understand why, one need only look at Washington's mixed record in Latin America. During Argentina's so-called Dirty War, for starters, as many as thirty thousand people were killed. Countless more were kidnapped and tortured. Roving "death squads" targeted politicos with ties to socialism and other left-leaning Peronist and Montoneros movements. In the final accounting, declassified cables later revealed that American officials not only seemed aware of the coming military coup but also understood that these Argentine officers were preparing to commit widespread abuses with "unprecedented severity." As thousands "disappeared," the junta's new foreign minister, César Guzzetti, communicated with the U.S. secretary of state, Henry Kissinger, telling him that his government was now going after "the terrorists." Kissinger, aware of growing public scrutiny of ongoing human rights abuses and increasing U.S. congressional pressure, responded by saying, "If there are things that have to be done, you should do them quickly. But you should get back quickly to normal procedures."

Then, on July 9 of that same year, 1976, Kissinger's top aide to Latin America, Harry Shlaudeman, was decidedly more direct, telling Kissinger that the Argentine tactics were akin to the "Chilean method," a reference to the Chilean dictator Augusto Pinochet. "That is, to terrorize the opposition—even killing priests and nuns and others." Kissinger twice urged Guzzetti to "get it over quickly," according to declassified documents. "The quicker you succeed the better." Guzzetti reportedly interpreted those comments as a "green light," even if subsequent cables by Shlaudeman suggested that Guzzetti "heard only what he wanted to hear."

Meanwhile, the junta's brutal tactics continued, as did the shutting down of the country's national legislature and the banning of trade unions, as well as a broad censorship to stamp out leftist guerrilla activity and other political challengers. When the extent of atrocities were ultimately revealed, they became both a national disgrace and a lasting stain on American foreign policy while also deepening the distrust between the two nations. Generations later, in 2016, Argentina's recently elected president, Mauricio Macri, would request a declassification of documents to both clarify events and gauge America's role in them. The timing of the request, however, was remarkable. At virtually the same moment that Argentina was asking Washington what it knew about one of its most traumatic periods in modern history, it was also suffering double-digit inflation, which many in the country blamed on the U.S.-based International Monetary Fund. Again, almost simultaneously, Beijing was now offering the cash-strapped Andean nation new lines of credit, along with plans for a Chinese ground station. With record droughts having ravaged much of the country, many were worried about yet another default. The country had already defaulted nine times, despite debt restructurings, leading to financial crisis after financial crisis. Buenos Aires was resultingly more than open to investments, and had found a willing partner in the East. By 2023, in fact, it had paid off part of its $2.7 billion IMF debt with Chinese renminbi. Argentine citizens were now able to open savings and checking accounts in the Chinese currency. High-profile Chinese infrastructure projects, meanwhile, continued to dot the countryside, including dams and port projects at the country's southern tip, near Antarctica. In Las Lajas, the Chinese ground station construction had been tantamount to a "four- or five-year bonanza," the mayor of Las Lajas, Maria Espinosa, would later tell me during my visit to the town.

"When the Chinese came, we had work," she explained. "All the rental spaces that we had in the town were 100 percent occupied." One of China's largest state-owned enterprises, a Beijing-based contractor called China Harbour Engineering Company, had plunked down hotel rentals for at least two years. The company, having become a subsidiary of China Communications Construction Company, was instrumental in China's ambitious Belt and Road Initiative—a massive global infrastructure project whose Chinese leadership was accused of leveraging debt to extract concessions from partnering governments.

But from the ground level, there were fresh insights that could be gleaned.

During construction of the station, for example, a member of a local government group named Juan Carlos Benavides had helped Chinese workers navigate the administrative tangles of their adopted country.

"They were quite private," he told me of the men, who included some two dozen Chinese military officials. Quiet at first, Juan explained, after a while, the officers warmed up.

"We didn't speak English, but we could understand and interpret some things," he explained. "They spoke sometimes in English, sometimes in Chinese, a mixture of English and Chinese that we didn't understand much.

"But what really caught their attention was how we lived," he added. "We have lots of free time. Plus, we Argentines are all barbecue and drink wine." On weekends, "we made them drink *fernet,*" he said, which is essentially just an Italian *amaro* liqueur mixed with Coca-Cola and served over ice. "They finished all of it."

Over time, however, other differences would prove more stark.

"Argentina is the country furthest from China," explained Lei Hao, an engineer on site. "We have different cultures and customs." Known for its powerful trade unions, for example, Argentina commonly experiences worker strikes that lead to canceled flights, stalled subway lines, and broader work stoppages. "There were protests and other things for labor issues. And the Chinese went crazy," Juan explained. "They could not conceive of a worker refusing to work."

In truth, labor strikes in China had become more common, with 434 reported factory strikes in 2023, compared with just 37 in 2022. And yet those numbers reflected merely reported strikes, which likely did not include military-run projects, like the Chinese ground station in Las Lajas.

When I arrived, the sheer size of the facility, and the rugged sweep of the surrounding highlands, had a certain rawness and immensity. Remote, dotted with security cameras, and adorned with razor wire, an eight-foot-high chain-link fence surrounds the station, where a thirty-five-meter white antenna punctuates the windswept landscape. Inside, a metal plaque depicting the Great Wall hangs in the lobby. Beyond it, scientists and engineers clad in blue lab coats work banks of computers, relaying much of their data back to receivers in China. The country, meanwhile, was also developing inter-satellite laser link systems using something called free-space optics, which afforded high-speed data transmission between satellites in low Earth orbit and reduced the overall reliance on ground stations. Advanced sensor technologies that made use of frontier quantum concepts, like entanglement swapping for instantaneous data transfer, were also under way. Yet for now, such ground stations were crucial for any kind of substantial activity in space. The teams worked, slept, and ate all within the facility, before being rotated out for a fresh crop of dish operators, scientists, and other workers.

Those in Washington, meanwhile, had grown alarmed. American surveillance concerns centered in particular on a phenomenon called signals intelligence, in which such stations could employ their massive antennas and receivers to intercept electronic signals, including satellite communications, cellular signals, radio broadcasts, radar emissions, and other types of electromagnetic signals. Geographic proximity to the intended targets mattered. The closer the station was, the more effective the collection. Of course, China has been operating what appeared to be an electronic eavesdropping station much closer to the United States since at least 2019, after Beijing formally signed a deal with Cuba in a bid to build overseas intelligence-gathering abilities. In June 2023, the National Security Council spokesman John Kirby said U.S. authorities were "well aware of—and have spoken many times to—the People's Republic of China's efforts to invest in infrastructure around the world that may have military purposes, including in this hemisphere."

Those concerns came to a head in 2023 when a Chinese spy balloon drifted over the entirety of the contiguous United States, only to be shot down off the coast of South Carolina. Intelligence gathering, be it personal, economic, or national security related, was a growing concern, particularly as trade policies between Beijing and Washington grew more hostile. Those concerns were now embodied in this massive new Chinese station. It was a symbol not just of China's growing ambitions but also of its willingness and ability to directly engage with international partners to achieve its goals. It was also just the kind of effort that those in Beijing had relied on people like Xue, and intermediaries like Stanislav, to forge.

Washington, meanwhile, slowly seemed to be waking up to the new reality. Its long-dormant ambitions in space were again stirring.

· CHAPTER 7 ·

# ARTEMIS

It was late afternoon in the summer of 2022 when a cream-and-yellow-colored microbus appeared on Florida's east coast. The guttural sputter of its air-cooled four-cylinder engine rattled through the humid air long before its distinctive, eyelike headlights could be seen against those distant heat waves. As it meandered over an Indian River causeway, one could be forgiven for thinking it wasn't an Artemis mission that thousands of spectators had traveled there to see, but rather her twin brother Apollo in the summer of 1969.

Half a century after Apollo 17 astronauts made humanity's last lunar journey, NASA was again preparing to return humans to the Moon. A diplomatic offensive was part of the strategy, known as the Artemis Accords, which sought to corral a shared commitment to norms and rules in space among dozens of nations. As for the Artemis missions, as commercial companies like Intuitive Machines and Firefly Aerospace scouted landing sites and built infrastructure, the idea (at least then) was also to establish a small space station called Gateway in orbit around the Moon before eventually erecting a base camp—if not several base camps—on the surface itself. The Moon was being positioned as a stepping stone for Mars, where ample sunlight, atmosphere, temperature, more suitable gravity, day-night cycles, presence of water, and abundance of $CO_2$ and other natural resources made it decidedly more attractive, even if a growing body of radiation-focused research suggested humans would struggle for longevity. Still, Mars was the goal. China had a crewed Mars expedition loosely set for 2050, just as Musk planned initial Starship missions there within the decade. But there were steps that had to be taken before any of that could happen. And in this race for influence, NASA's Artemis I—the first real test of the agency's Space Launch System (SLS) that aimed to return humans to the Moon—was about to take flight.

As the Sun dipped below the horizon, camper vans and pop-up tents clustered in designated areas across the launch-scape. I had rented a truck and driven it onto Merritt Island, a barrier island on Florida's eastern shore, also home to the Kennedy Space Center. There, telephoto lenses perched along rows of tripods, aimed at the distant rocket and iconic silhouette of NASA's Vehicle Assembly Building, originally constructed to assemble the Saturn V rocket for the Apollo program. The agency had since developed its own launch system, with Boeing serving as the lead contractor for the core stage, upper stage, and flight avionics, contracting to SpaceX and Blue Origin for separate Moon landers—seemingly a mixed product of NASA's old and new ways of doing business. SLS and the Orion capsule were meant to first deliver crews into lunar orbit. Gateway would serve as the staging point for human Moon landings, with commercial landers meant to take astronauts the rest of the way. It was a curious setup, and one that included an extra, and highly controversial, step in the construction of Gateway.

"I will be direct," explained the former NASA administrator Mike Griffin. "In my judgment, the Artemis program is excessively complex, unrealistically priced, compromises crew safety, poses very high mission risk of completion, and is highly unlikely to be completed in a timely manner even if successful." Griffin didn't appear to be wrong. But it was the commercial partners aspect that he generally opposed. And yet by the end of 2025, NASA would announce it was opening its human lunar lander contracts to new competition following delays with SpaceX.

Still, there seemed no denying that there were major problems with the legacy approach, which companies like SpaceX and Blue Origin were meant to resolve. Envisioned more than a decade earlier when President Obama signed the NASA Authorization Act in 2010, which called for SLS to be ready in six years, it was by now grossly over budget and well behind schedule—a curious thing given much of its hardware wasn't even new. Cobbled together from retired space shuttles, the vehicle included such old components as a booster aft skirt first used in 1984. The hope had been that by using refurbished components, it would be cheaper. In reality, the redesign might have proved more time and cost consuming than building from scratch. Part of the problem was that the "complexity of developing, updating, and integrating new systems along with heritage components proved to be much greater than anticipated," explained NASA's inspector general Paul Martin. But the real head-scratcher, again, was the fact that SLS was a single-use rocket. In an era where the first stages of Falcon 9s routinely made round trips, and where the far larger Starship was also being designed for reusability, expendable rocketry seemed like a relic of the past. And it certainly wasn't the vehicle that the Chinese had chosen to emulate. They were clearly modeling their next-generation vehicles on reusable systems, not throwaway ones.

Plus, few seemed certain the system was even up for the mission at hand. NASA's Office of Inspector General (OIG) found that Boeing, responsible for the core stage of the SLS, and whose contract swelled from $962 million to more than $2 billion through 2025, was confronting what appeared to be serious quality-control issues. "According to DCMA officials, Boeing's process for addressing contractual noncompliance has been ineffective, and the company has generally been nonresponsive in taking corrective actions when the same quality control issues reoccur," according to the NASA watchdog report.

The SLS Block 1B version, which featured an upgraded upper stage meant for future Artemis missions, had been in development since 2014, which, of course, was a decidedly different era. At that time, Boeing commanded a majority of the funding for NASA's Commercial Crew Program, and nearly twice that of SpaceX. But times had changed, especially given the OIG report had pointed to evidence of an insufficiently trained workforce, citing welds that failed to meet specifications "due to Boeing's inexperienced technicians and inadequate work order planning and supervision."

These, it seemed, could be the early warnings of deeper issues, especially when in 2024 Boeing's Starliner spacecraft was compelled to leave its crew aboard the International Space Station. During that flight, the astronaut Butch Wilmore had to take manual control of the vehicle, which he later explained to the journalist Eric Berger.

"As Starliner's thrusters failed," Berger reported, "Wilmore lost the ability to move the spacecraft in the direction he wanted to go," as concerns mounted as to whether the craft could either dock or return safely to Earth. Starliner would ultimately return safely, but would do so without its astronauts, an unmistakable showing of the sheer lack of confidence in the craft. In the interim, Wilmore and his fellow astronaut Sunita Williams's eight-day mission would be extended to roughly nine months aboard the ISS, with a SpaceX Dragon capsule—Boeing's direct competitor—ultimately responsible for bringing them home. Before launch, Erin Faville, president of a NASA contractor called ValveTech, had urged Boeing to suspend its upcoming Starliner launch "due to the risk of a disaster." In fact, the Atlas V rocket that carried the Starliner capsule had to be rolled back to its hangar to swap out a pressure valve after a "buzzing" sound before launch alerted engineers that the valve not only may be past its prime but, according to Faville, was not suitably qualified for human spaceflight. The issue seemed to stem from a 2011 dispute between Pratt & Whitney Rocketdyne and ValveTech surrounding NDAs and trade secrets of the valve technology, but a Boeing spokesperson promptly refuted those claims, saying the valves "meet all NASA and Boeing requirements."

Still, a full decade into development, concerns over leaks, parachute problems, and hundreds of feet of flammable tape inexplicably used as insulation on wiring inside the capsule itself were mounting. The chair of a NASA safety panel, Patricia Sanders, was calling for the agency to take an independent "deep look" at the various issues plaguing Starliner, noting that it "should not be flown until safety risks can either be mitigated or accepted, eyes wide open, with an appropriately compelling technical rationale." Both Boeing and SpaceX had been awarded six flights each by the agency as part of NASA's Commercial Crew Program. And yet SpaceX would complete all of its flights before Boeing could successfully complete its first, a clear sign of SpaceX's dominance in launch, satellite, and human spaceflight.

No other company could match it. Dragon capsule was by then the only reliable way to get to the ISS (without relying on the Russians). The company's Falcon 9s and Falcon Heavies flew sometimes multiple times per week. And the Starlink constellation was by far the world's biggest. There was also simply no comparison to the nascent Starship, a version of which had been baked in Artemis plans to return astronauts to the Moon. SpaceX seemed almost too big to fail, a curious development in the race with Beijing given Musk's extensive business dealings with China, largely through his electric car company. Tesla, whose shareholders in 2025 approved a record pay package that could provide Musk with roughly $1 trillion in company stock over the coming decade, also maintains a Gigafactory in Shanghai that has accounted for more than half of the company's global sales. With access to top Chinese leadership, including the Chinese premier, Li Qiang, who as former party secretary for Shanghai praised Tesla as a "successful model" for U.S.-China collaboration, such dealings prompted those like Senator Mark Warner (D-Va.) to question the SpaceX founder's "deep financial exposure to China."

"I think I understand China well," Musk said in 2023, speaking remotely at the All-In Summit in Los Angeles. "I've been there many times. I've met with senior leadership at many levels of China for many years. And so I think I've got a pretty good understanding as an outsider of China. And Tesla has been very successful domestically in China." The bigger question, it seemed, was whether those ties posed an increasing conflict of interest, given SpaceX's expanding role in U.S. national security and Musk's ambition to morph the car company into a powerhouse of both data and robotics.

For now, however, questions surrounding Musk's dealings with China would remain unresolved. For the moment, however, attention had shifted to the Cape, where anticipation was building around the launch of Artemis 1, and the partnerships and negotiations behind it, sealed as they were with a touch of space-age Americana.

· CHAPTER 8 ·

# SELLING THE MOON

Mike Gold seemed like a Beltway guy. The former NASA associate administrator had worked for Maxar Technologies and Bigelow Aerospace, often operating out of Washington, D.C. A lawyer by trade, by the time we spoke, he was part of Redwire's executive leadership team, intent on leveraging technologies to tap into existing markets on Earth. Communication networks, weather forecasting, or the development of pharmaceuticals and 3D-printed organs were all prime examples. In fact, Redwire's first-ever human knee meniscus and cardiac tissue had been bio-printed aboard the ISS, which had been marked for retirement by the end of 2030, giving way to a bevy of commercial stations, including Orbital Reef by Blue Origin and Sierra Space, Starlab by Nanoracks, Voyager Space, as well as Axiom Station by Kam's Axiom Space, among others. Microgravity, in fact, is uniquely advantageous for experimentation. In a biological sense, tissue cultures tend to collapse in gravity, whereas in microgravity they seem to mimic more closely how cells develop in the body. There, each layer can be printed with a gel-like scaffolding that makes it easier to replicate the three-dimensional structure of organs. A similar approach can be used in the development of cancer-fighting drugs, allowing for printing in remarkably reliable ways that can help identify how certain drugs bind to tumors.

"Microgravity drug development is really low-hanging fruit relative to space commercialization," Mike explained, something that Janet Kavandi, then Sierra's president, had also echoed. It was a core part of answering the "what can we build in space that pays the bills?" question, she told me one afternoon in the company's Louisville, Colorado, headquarters. Create a product that can only be produced in space, or made in a vastly superior way—the thinking went—and market demand would follow.

And yet in the years before he joined Redwire, Mike had been faced with a considerably different task, one he admitted he had been hesitant to take on: With the Moon as a rallying point, America's renewed space efforts needed partners to build momentum.

That's where his love of *Star Trek* came in.

"As a *Star Trek* fan, this was literally the birth of the United Federation of Planets," he told me, a reference to a fictional interstellar alliance of planets and species from the show. The Artemis Accords had, in effect, been inspired by *Star Trek*'s Khitomer Accords, a reconciliation between the federation and Klingons. And in a moment seemingly ripped from those old VHS tapes, Mike would travel the world to rally support with that as its touchstone.

Turns out, it wouldn't be easy.

"It is extraordinarily difficult to gather the consensus foreign agreement, not only with the initial group where we're dealing with eight different space agencies, but we also had eight different foreign ministries that were working with us," he said. "So even though initially, there were eight countries, there were actually sixteen different entities involved."

To get everyone on board, he needed a narrative and a mission. Unlike NASA's plans to send a robotic spacecraft to lasso an asteroid, the Moon "was widely accepted as a logical next step for human exploration." It was also "popular among the international community," he explained. Plus, international agreements tended to be harder to unwind than domestic policy agendas. If partnership could be forged across borders, that might give Artemis some staying power across successive administrations. "We talk about not only the importance of leading in technology, but policy as well."

"The first people [there] are going to get to set those norms and behaviors."

Clearly, he was referencing Beijing.

"I think there is a historic opportunity to work with the Chinese and ILRS [International Lunar Research Station] nations to prevent space policy and law from bifurcating along Artemis and ILRS lines," Mike added. "We need to get [back to the Moon] first. But where we can work together for mutual benefit, we should also do so."

By the fall of 2020, those efforts had forged an initial framework. Seven countries—Australia, Canada, Japan, Luxembourg, Italy, the United Arab Emirates, and the U.K.—had all joined the United States in signing. Dozens more would follow in the years ahead, an effort Gene Roddenberry himself might've admired, with the intention of staking out the rules of this new space era before anyone else could. Plus, a closer look could reveal what was really at stake for nations, companies, and their looming claims on the Moon.

In a 2019 interview with *Politico,* NASA administrator Jim Bridenstine had said, "We are building an architecture where we will have access to any part of the Moon, at any time." It was a bold statement. But it also, in a way, highlighted the very complexities of future claims. The 1967 Outer Space Treaty, a bedrock agreement signed by the United States, the U.K., and the Soviet Union, as well as 114 other nations (including China), explicitly prohibited ownership of celestial bodies, which included the Moon, and presumably the kind of commercial access (and ownership) that might spawn the makings of a real lunar market. Created two years before the Apollo Moon landing, and at the height of the Cold War, to ensure space remained a place of peaceful exploration and shared scientific progress, the treaty suddenly seemed like an impediment to growth. How could a viable space economy develop without the kind of ownership that this space law was specifically designed to prevent? With rising commercial interests, the Outer Space Treaty appeared increasingly open to interpretation.

And by 2015, the Obama team had sought to clarify it, signing the Commercial Space Launch Competitiveness Act that made it legal for U.S. privateers to own and sell the space resources they extracted while offering liability protections for commercial actors operating in orbit, including space tourism. Two years later, the

Trump administration effectively doubled down, clarifying that the U.S. position viewed space not as a "global commons" but as a "legally and physically unique domain of human activity."

"Americans should have the right," the executive order read, "to engage in commercial exploration, recovery, and use of resources in outer space, consistent with applicable law." Both administrations were making the case that the Outer Space Treaty had loopholes that allowed citizens to have de facto claim to the celestial places they mined, without technically claiming those regions as their own. An analogy in U.S. Code, in fact, likened celestial mining to "harvesting fish from the sea" in international waters. The Moon was to be made open for business, even as American rivals protested.

The first comments came from Moscow.

Roscosmos's director general, Dmitry Rogozin, called Artemis "a big political project," resembling a new kind of NATO in space, of which Russia would not take any part. In fact, Beijing and Moscow were actively courting partners of their own, perhaps in tacit recognition that a unified framework in space had yet to emerge. The world in many ways was now pairing off between East and West, much as it had in the old Cold War, albeit without an agreed-upon set of standards to guide it. More than seventy nations now had space programs. And while governments were still responsible for overseeing their private sectors, getting satellites and spacecraft into orbit no longer required nation-states, or even the budgets of major contractors, with student-built CubeSats tallying as little as a few thousand dollars to reach orbit. Space was wide open. And yet the moves also seemed tantamount to a kind of finders-keepers policy, especially in commercial mining. Foreseeing future tensions, the accords included "safety zones" between lunar bases, which were neutral areas designed to prevent conflict, especially over the extraction of water ice. Still, without a global set of standards, the risks of dispute and conflict were growing, especially when it came to mounting militarization and emerging claims of territorial sovereignty.

For many in Beijing, these developments were more than just

strategic challenges. They struck at deeper national concerns. Future explorations were seen not just through the lens of security and commerce, but as part of a broader effort to reclaim China's historical narrative—one that had long been shaped by imperial histories and colonial infringements—with a renewed sense of its own national destiny.

· CHAPTER 9 ·

# ECHOES OF CHINA'S RISE

They did not want him having sex. And with the flick of an imperial blade, they made certain he would not. Zheng He, thought to be an unusually tall man widely dubbed China's greatest admiral and explorer, might have made an attractive father had he not been gelded as a boy. But as a young Muslim, thought to be the great-grandson of a rival Mongol warrior, Zheng—or Ma He, the Chinese version of Muhammad, as he was known—was instead to be subjugated. The Ming dynasty had begun. Its need for soldiers and servants was growing. Conquered peoples were a ready source, castration a common practice. Zheng, his attackers likely presumed, having been made a eunuch and thrust into servitude, would produce no rival sons or daughters, his genetic lineage lost forever.

In a way, they were wrong; his legacy passed down through the generations, not by blood, but by history. Nations, to be sure, need their heroes, especially in space. The Soviets had Gagarin. America had Armstrong. And China seemed bent on creating its own age of Apollo, albeit with its own nod to the past. In that context, it is helpful to first consider Zheng, which Chinese leadership often does, including President Xi, who—among other instances—invoked his name in Paris during a March 2014 UNESCO speech:

> In the early 15th century, Zheng He, the famous navigator of China's Ming Dynasty, made seven expeditions to the Western Seas, reaching many Southeast Asian countries and even Kenya on the east coast of Africa. These trips left behind

> many good stories of friendly exchanges between the people of China and countries along the route. . . . There were indeed conflicts, frictions, bewilderment and denial in this process. But the more dominant features of the period were learning, digestion, integration and innovation.

Now, he added, "the Chinese people are striving to fulfill the Chinese dream of the great renewal of the Chinese nation." Xi would use his name again three years later at the opening ceremony of the Belt and Road Forum for International Cooperation in Beijing, calling Zheng's prowess across the seas a "feat which still is remembered today." During that same speech, he also sought to placate concerns about China's rise by reinforcing notions that such pioneers "won their place in history not as conquerors with warships, guns or swords" but "as friendly emissaries leading camel caravans and sailing treasure-loaded ships."

China, of course, was then a much different place, as was much of the rest of the world. Europe was just emerging from the Dark Ages, whereas Ming imperials had sought a man like Zheng to buttress a growing empire, in which new frontiers of trade were being charted as seaborne commerce and shipbuilding exploded. By then, Chinese naval technology was virtually unmatched, with shipbuilders and navigators employing magnetic compasses, along with star charts, lateen sails with multiple masts, sternpost rudders, and double hulls, long before their European counterparts were capable of ferrying massive crews and cargoes in such distant voyages. Power and influence washed over the Chinese capital as the nation's port cities swelled as emporiums of raw goods and ideas, breathing life into the nation's economy. The last of the three golden ages had begun, with architecture, printing, literature, and medicine all flourishing. Wealthy merchants flocked to China's seaports. Foreign ambassadors marveled at its prestige. And a global system of tribute supported by Zheng's expeditions would cement Ming dominance and guide imperial building and expansion.

No longer a mere methodology for prehistoric survival, exploration was now a national priority, the state-funded seedlings of global influence, setting the stage for subsequent explorers like Magellan, Vespucci, and even Armstrong. Yet as wealth and prestige amassed, so too grew fears of how much could be lost. To protect what they had built, and wary of the burgeoning threat of Mongol invaders, legions of fifteenth-century Chinese workers began cobbling earth and stone to complete a section of the Great Wall known as the Badaling stretch so as to guard Beijing and its imperial palace. It remains the wall's best-known and best-preserved section.

And yet curiously, upon completion, a metamorphosis in the way global trade and exploration were valued within Chinese society simultaneously seemed to be taking hold. Neo-Confucianist thought, with its comparatively more inward and at times ethnocentric focus, was again fashionable as China's mercantile ethos and penchant for expansionism seemed to erode. A more preservationist politics would emerge, giving power to those who criticized Zheng's expeditions for the huge cost and drain on resources that might otherwise have been directed toward domestic needs. They might have been right. These were enormously expensive expeditions, whose return in prestige provided only marginal benefit as they continued, perhaps much in the same way America's Apollo missions would come to be viewed after Apollo 11 as public interest waned. Yet for the Ming dynasty, the fifteenth century marked the beginnings of underlying issues that would later destabilize the empire—namely new restrictions on trade, growing bureaucratic corruption, and mounting fiscal strain that stemmed from public works programs and military campaigns that included conflicts against Mongol tribes to the north and Vietnamese resistance to the south, as well as a series of internal rebellions.

Following Zheng He's death, somewhere between 1433 and 1436, China would turn inward, reversing a period of expansive international engagement. The decision coincided with Europe's own age of exploration and a pivot period of burgeoning global trade, in which European explorers sought new trade routes, resources, and territories, ultimately gaining a degree of dominance in maritime trade as Europeans sought to colonize parts of Asia and the rest of the globe through expansive public-private partnerships. Having simultaneously begun to pull inward, while building walls over a perceived Mongol threat, leadership in China would begin to cede something far more profound: the shaping of the emerging world order. As China stressed internal stability, essentially fortifying itself against external threats, European powers pushed outward. Walls—unlike ships—are not built to trade and explore. Meanwhile, a growing number of European explorers disembarked in Chinese port cities. Soon after, powerful joint-stock enterprises—such as England's East India Company—would pursue their own commercial and strategic interests abroad, often with the strategic backing of their home governments. A hallmark of the era, new financial incentives and stock options now spurred entrepreneurial growth and drove an expansion of empires, perhaps not entirely unlike the beginnings of today's commercial age in space. What eventually followed in China became known as the century of humiliation, the result of infighting and technological inferiority to colonial powers and their commercial interests, as well as the illegal importing of opium to China, which resulted in widespread addiction and social and economic distress. By 1860, nearing the end of the Second Opium War, Britain's high commissioner to China, Lord Elgin, ordered his forces to destroy the Old Summer Palace as retribution for the slaughter and torture of British envoys in Beijing. Considered a masterpiece of architectural and landscape design, the palace, symbolizing China's dynastic authority, also held immense cultural, imperial, and spiritual significance.

Elgin almost certainly knew this. He wanted to send a message.

In ordering its destruction, Elgin wrote that he meant "to mark, by a solemn act of retribution, the horror and indignation . . . with which we were inspired by the perpetration of a great crime."

"What would the *Times* (newspaper) say of me," he reportedly told a French commander, "if I did not avenge its correspondent?"

But it was as clear a sign as any of China's relative weakness. At the war's conclusion, Beijing would be forced to accept two humiliating conditions: The first would be the loss of the territory of Hong Kong to British control—a loss that would extend for the next 156 years. The second would green-light British opium smugglers into

China, offering them—despite Chinese bans on opium's sale and consumption—special access to Chinese ports. By 1840, more than ten million Chinese people were thought to be opium addicts, fed primarily by illegal imports, which had helped tip the China–Great Britain trade imbalance in London's favor, much at the behest of that East India Company, which had established a veritable monopoly on opium cultivation in the neighboring Indian province of Bengal, before shipping the narcotic to Chinese ports. By the conclusion of the Second Opium War, that trade had only increased. (Today, China is rather curiously a key source of fentanyl smuggled into the United States via Mexico, and one can speculate about the degree of irony especially as is witnessed from Beijing.) In contemporary China, the Opium Wars are often taught as a cautionary tale, at times peeling back the pages of history to the comments of the British politician William Gladstone, who would later become prime minister, and his own denunciations of the conflict.

"A war more unjust in its origin, a war more calculated in its progress to cover this country with permanent disgrace I do not know, and have not read of," he then told Parliament. Since then, a common phrase in China, relating to that era, is *luo hou jiu yao ai da,* seen as a warning and loosely translated to mean "if you are backward, you will take a beating." President Xi repeatedly reminds his listeners of those infractions in public speeches, referencing his nation's past "humiliation and sorrow," the ruins of the old palace left noticeably unrepaired as a stark reminder that when China looked inward, it was exploited. When it looked outward, it thrived. And yet now, nearly six centuries after that famous Chinese admiral made his last voyage, leadership seemed intent on avoiding mistakes of the past.

By the start of the twenty-first century, China had made history, becoming only the third nation, after the United States and Russia, to send an astronaut on its own vessel into space.

"Frankly, China is moving at a breathtaking speed," said General Stephen Whiting, commander of U.S. Space Command, in a 2024 speech at the Thirty-Ninth Space Symposium. Though the speech largely centered on Chinese in-orbit intelligence, surveillance, and reconnaissance satellites, his remarks also came just weeks after the launch of the Queqiao-2, a key relay satellite meant to support the fourth phase of the Chinese Lunar Exploration Program.

The first three phases centered squarely on (1) orbital missions, (2) soft landings and lunar rovers, and (3) sample returns. But the fourth phase, in which the groundwork for a lunar research station was being laid, in many ways seemed like the bigger deal. Not only did it comprise an orbiter, return vehicle, lander, and ascent vehicle, using a docking maneuver similar to that of the old Apollo missions, but it also made use of a communications relay satellite that represented the beginnings of a nascent lunar communications network. Ge Ping, deputy director of China's Lunar Exploration and Space Engineering Center, called it a "key link" for a "communications hub."

And yet, for the moment, Beijing was treading cautiously, with its planned station meant to align (at least publicly) with the 1967 Outer Space Treaty, which stipulated that the "moon and other celestial bodies" were "not subject to national appropriation by claim of sovereignty, by means of use or occupation, or by any other means." And yet a review of Beijing's approach to international maritime law might be useful in predicting future Chinese actions. In 1982, China was among the first nations to sign on to the United Nations Convention on the Law of the Sea (UNCLOS). Back then, again, China was a different nation. The country was still in the early stages of economic reform initiated by Deng Xiaoping, and was still emerging from the relative international isolation that it had experienced under Mao Zedong. Its military, though it included a massive standing army, was hardly a modern force. It lacked advanced weaponry, and its air force and navy were considered underdeveloped. As such, Beijing had more modest territorial ambitions. Hong Kong was still under British sovereignty. And the concept of the nine-dash line, which outlined China's expansive maritime claims in the contested South China Sea, was not as prominent a part of the nation's foreign policy as it would be. China of 1982 had signed on to the UNCLOS presumably to show a willingness to play nice when it came to international norms and standards. But over the years, as its power swelled, replete with astonishing economic growth, military modernization, and the kind of technological advancement that began to establish itself as a global leader, Beijing appeared less concerned with keeping a low profile. That may help explain why, when the Permanent Court of Arbitration in The Hague concluded in 2016 that China's claim to nearly the entire South China Sea—particularly the Spratly Islands and Scarborough Shoal (including Huangyan Island), which are both rich in oil and gas—was effectively baseless, Chinese leadership essentially ignored the ruling, calling it "null and void." When it came to such territorial claims, not only had China seemed to place itself above international law, but it also began exercising something it called veto jurisdiction, blocking other countries' ability to survey, explore, and make use of marine resources in these disputed waters. Similar policies were applied to the Diaoyu Islands, which China claims are "under its jurisdiction," but are also claimed by Japan, with patrol vessels from each nation surrounding that small scattering of uninhabited islands. Those decisions fueled growing questions about Chinese intentions on the Moon, in Earth's orbits, and indeed across the solar system.

"I know what China has done on the face of the Earth, for example, where the Spratly Islands, they suddenly take over a part of the South China Sea and say, 'this is ours, you stay out,'" NASA administrator Bill Nelson explained to the NPR journalist Scott Detrow in 2024. "I don't want them to get to the South Pole . . . and say, 'this is ours. You stay out.'" To date, China—like the United States—considers the mining and exploitation of space resources permissible, provided it's done in accordance with the Outer Space Treaty, and has even emphasized collaboration, having signed scores of international agreements over the past decade, including a 2016 memorandum of understanding with the UN Office for Outer Space Affairs to offer up the Tiangong space station to all UN member states. Even its prospective outpost on the Moon is referred to as the International Lunar Research Station. And yet it is also not the powerful spacefaring nation it hopes to be, a position that might be likened to its maritime position in 1982, when leadership signed treaties and showed far greater willingness to conform to international standards. Will that spirit of cooperation persist as China's space presence grows stronger? Or will Beijing again become more assertive and willing to bypass international law for the sake of its own interest as it did in the South China Sea? For context, we should consider the words of Ye Peijian, China's own lunar exploration program chief designer:

> The universe is an ocean, the Moon is the Diaoyu Islands. Mars is Huangyan Island. If we don't go there now even though we're capable of doing so, then we will be blamed by our descendants. If others go there, then they will take over, and you won't be able to go even if you want to. This is reason enough.

A few years later, more clues started to emerge.

In 2023, a Chinese company called Hong Kong Aerospace Technology Group signed a deal with Djibouti to erect a rocket launch port in what constituted the first such Chinese facility in an overseas territory. One practical reason for the move was geographic advantage, given its closer location to the equator, which afforded rotational speed boost from Earth. In other words, Djibouti allowed for more fuel-efficient launches than from higher-latitude facilities in China. And yet there might be another reason why it was chosen. The small African nation had never signed the Outer Space Treaty, thus offering a potential end run around existing legal frameworks. As Benjamin Silverstein, lead research analyst for the Space Project at the Carnegie Endowment for International Peace, wrote, "Djibouti's lack of experience with any form of space law does little to raise hopes for effective oversight of activities originating from its soil." And, he posited, "by building spaceports overseas, Beijing wants to flout the current rules, and write its own." Yet legal maneuvering seemed like just the beginning of an ambitious plan, far more comprehensive than anything seriously outlined in the West. Chinese space officials were now, in fact, not so quietly laying the groundwork for an audacious vision that extended to the outer edges of the solar system. That road map, Tiangong Kaiwu, named after the pioneering work of the Ming dynasty scientist Song Yingxing, identified strategic mineral resources, placements of transport and supply nodes, and locales for harvesting water-ice fuel, in what amounted to an extraterrestrial supply tributary, perhaps not wholly unlike the focus of the Zheng missions of old.

Goals were laid out for 2035, 2050, 2075, and 2100, and were meant to "promote the development and utilization of space resources in China to achieve leapfrogging-style development," said Wang Wei, a scientist affiliated with the China Aerospace Science and Technology Corporation (CASC), China's main space contractor. "Just like the miracles created in the Great Age of Navigation," he explained, "the 'great space age' with the use of space resources will . . . create the next miracles in the history of human development and bring new prosperity to human civilization." Many of the key areas focused on Lagrange points, where gravity from different celestial bodies can effectively cancel each other out. Like eddies in a river, the net result allows objects to stay in relatively stable positions in space with limited expenditures of fuel. As infrastructure developed, the use of Lagrange points as staging areas seemed to be especially useful, albeit limited, providing strategic advantages to those who controlled these locations in space. Chinese space plans, however, would get much more specific. During a Chinese Society of Astronautics meeting in Beijing, for instance, Wang Wei outlined a road map for celestial resource development. Using both the Moon and Lagrange points as stepping stones, it endeavored to explore, mine, and set up supply lines across the entire solar system by 2100, which involved space plans and reusable launch vehicles, as well as the construction of a nuclear-powered shuttle by 2045. The propulsion reactor for the latter had by 2022 already passed its initial performance evaluations, which state media claimed to be a hundred times more powerful than a similar reactor NASA had envisioned for its own lunar base.

Meanwhile, Yang Mengfei, a top space official also from CASC, was urging for still more speed.

"At present, the United States, Europe, and Japan have proposed relevant plans for Earth-moon space infrastructure, but they have not yet entered the stage of on-orbit construction," Yang said, according to a translated CASC statement. The current moment represented a fleeting chance "to seize the opportunity and lead the Earth-moon space industrial market," he added. "It will have a great impact and far-reaching significance."

Fueling concerns was an October 31, 2021, edition of *Science and Technology Daily*, the official newspaper of China's Ministry of Science and Technology, in which a senior CASC official, Bao Weimin, described Chinese efforts to establish a special economic zone between the Moon and Earth that could yield as much as $10 trillion in annual revenue by 2050.

Wang Wei, then lead scientist at CASC, also identified 122 near-Earth asteroids he considered viable for mining. Clearly, Chinese leadership saw intrinsic value and was investing accordingly. Chinese astronauts, or "taikonauts," meanwhile, continued to blast off from the Jiuquan Satellite Launch Center, a major spaceport in the country's northwest, which served as the nation's first satellite launch facility. Fittingly perhaps, it was there, on the edge of the Gobi Desert, that China would first emerge as a global space power. And like U.S. leaders with the Apollo Moon landing in 1969, its senior officials seemed to recognize the galvanizing power of a good story.

· CHAPTER 10 ·

# FIRST TAIKONAUT

It was a blue-sky morning on the edge of the Gobi Desert when a thirty-eight-year-old Chinese fighter pilot named Yang Liwei strapped into a Shenzou spacecraft, translated as Divine Vessel. It was October 15, 2003. And at first blush, the vehicle looked remarkably similar to Russian designs, albeit slightly heavier. Moscow and Beijing had indeed been collaborating when it came to space for considerably longer than many in Washington had realized. Just two years earlier, during a NASA forum convened at George Washington University on the condition of anonymity, one participant noted that he had been "struck [by] the technology flux of systems between Russia and China," which had manifested itself in ballistic missile technology, satellite development, and rocket and capsule design.

"The Shenzou spacecraft shows how far things can go without us noticing," the speaker noted, according to documents posted on the university site. "This could be an example of many other transfers going on between Russia and China without the United States noticing." In 1992, Moscow and Beijing would, in fact, sign their first major space deal, focused primarily on exchanges of research and technological expertise. Over time, as the dealings progressed, upper-stage launch vehicles, a joint missile early warning system, space debris monitoring, ground stations for dual military-and-civilian coordination, and even a lunar lander were on offer.

Fast-forward to 2003, and a global demonstration of that collaboration was about to take shape. There were, in fact, few moments in the early history of China's space program that would be as notable as what Yang had now set out to accomplish. At a restricted launchpad in the country's northwestern Gansu province, the son of a teacher and accountant, and father to an eight-year-old son, was preparing to make history. More than forty years after Yuri Gagarin first circled Earth, Lieutenant Colonel Yang was about to make his country only the third nation to ever send a person into space, staring down the barrel of a twenty-one-hour flight in a capsule that was expected to make fourteen orbits around Earth.

"You carry the dreams of our nation into space with you," President Hu Jintao told Yang just minutes before takeoff.

At 9:00 a.m. local time, the engines of that Long March CZ-2F rocket ignited, carrying Yang into clear skies. As the vehicle rumbled skyward, eventually tipping on its side as it spiraled into orbit, onlookers in attendance roared in celebration. Rows of balloon-wielding children had come to see him off. Roughly half an hour into the flight, Yang radioed back to mission control in Jiuquan. "I feel good and my conditions are normal," he said. That ease, as would be later disclosed, was apparently short-lived, however. Violent vibrations were by now rattling through the craft as Yang became disoriented.

"I felt my internal organs were upside down," he explained in a subsequent state-media-produced documentary. "I suddenly lost any sense of direction."

Entering China's crewed spaceflight program amid a pool of more than fifteen hundred candidates, he and eleven other taikonauts had dedicated five years to the program. Much unlike the fanfare that surrounded Project Mercury, which put the first Americans in space and manned six flights between 1961 and 1963, Yang's name would be revealed just a single day before launch. Similar to Gagarin, he would be promoted to colonel just prior to the flight. But now, as the craft gathered speed, Yang had "a feeling I never had in my training." There was no live television coverage of the event. And so in a carefully crafted disclosure, footage of the flight later portrayed him calmly eating in microgravity, floating what appeared to be a candy bar into his mouth.

It was only years later that the dramatics of the event would be revealed. Not surprisingly, the event had been meticulously choreographed. After all, China's space program was still working to shed the shadow of what many grimly recall as its "dark decade." Throughout much of the 1990s, the nation's space agency had been plagued by high-profile incidents that rattled both public confidence and internal morale. The program had "fallen into a trough with continuous failures," reflected Chen Zhenguan, a former research fellow of launch vehicle technology. It was a time marked by setbacks, when ambition often outpaced capability. Nevertheless, there was still opportunity here. Back then, with NASA's decision to restrict commercial payloads in the aftermath of the 1986 *Challenger* disaster, U.S. satellite companies had been looking elsewhere to launch their technologies. China offered comparatively bargain deals for rides atop its newly built Long March rockets, which undercut European competitors, sometimes by as much as half. Meanwhile, America's fledgling private satellite market was just getting going. Chinese facilities were not up to U.S. standards. But if they generally proved functional, that might be enough. Traveling out to evaluate the facilities, one Intelsat representative reportedly described the Xichang launch center as "primitive but workable."

In 1996, however, disaster struck. On February 15, the engines on the Long March 3B rocket carrying an Intelsat satellite ignited. At liftoff, a group of engineers moved outside for a better view. "I got out, turned and ran around the building to my best viewing spot, in time to see the mountain lit from behind, hear the startling rumble and see the rocket emerge," wrote one U.S. engineer working at the facility in his diary, as reported by the space journalist Anatoly Zak. "But instead of rising vertically for nine seconds and several thousand feet, I saw it traveling horizontally, accelerating as it progressed down the valley, only a few hundred feet off the ground. 'Wrong way!' I yelled, and for the next few seconds I was frozen in my tracks."

The 426-ton rocket then plunged into a neighboring hillside, producing the "biggest explosion of my life," the diary noted, as reported by Zak. It killed at least six people, proving both a national embarrassment and a hallmark of events to come. Even as the Chinese space program improved and grew more sophisticated, rocket explosions in populated areas would persist. As late as 2024, video published on Chinese social media appeared to show a booster fall into a populated region of Guiding County, Qiandongnan Prefecture, in Guizhou province—believed to be part of the Long March 2C rocket, which had blasted off earlier from Xichang Satellite Launch Center on June 22. (Inland launch centers, as opposed to coastal ones, could prove dangerous to local populations, though back then tended to be harder to surveil.)

It was, in fact, that 1996 catastrophe that prompted Liu Jiyuan, former general manager at CASC, to effectively realize that "it was more complicated than I thought for China to enter the global commercial satellite launch service market," a realization that led to reforms and greater scrutiny over safety protocols. Still, questions remained about Chinese capacities, especially as the nation endeavored to put a human being on top of one of those rockets. China needed a win. Another disaster could spell serious setbacks. And yet part of the strategy was also about better controlling the narrative.

As for Yang in his Shenzou craft, the mission was so far holding together. As the vehicle circled Earth, news of the coming accomplishment filtered back home. President Hu had already called it a "significant, historic step by the Chinese people in scaling the peak of the world's science and technology." But Yang, alone and crammed inside a tiny capsule, still had to get back. It was just after 6:00 a.m. EDT as four tracking ships in the Pacific, Indian, and southern Atlantic Oceans were monitoring. As the craft descended over Africa and soon Pakistan, it reengaged with Earth's atmosphere, and the shaking resumed.

As is customary during reentry, the Shenzou appeared as a veritable fireball.

"Everything turned red outside," he recalled. But, he remarked, there was something else: a distinct high-pitched sound. One of the windows, designed to withstand the external atmospheric pressure and thermal stresses, was apparently cracking. As the heat continued to build, he could see the cracks more clearly. "It was terrifying, really terrifying. No one had told me this could happen. I had no idea. I thought the windows were going to shatter. . . . I was strapped tight to the seat. I could barely move when the window on one side broke [though not fully]. Then the other side began cracking too." And yet as the seconds and minutes ticked by, the windows, although cracked, seemed to be holding. "After that, I calmed down," he said. "On one side, the cracks weren't too bad. And it was unlikely that both sides would fail."

He was now passing over Tibet before reaching Inner Mongolia, where the vehicle's heat shield jettisoned and its parachute opened some twenty miles above Earth, with the craft's retro-rockets firing as it touched down. Yang would emerge a national hero, labeled as such by military leadership, who touted his "heroic achievement." Saturation coverage by the state-run Chinese media quickly followed. Yang toured the country. It had been similar for Gagarin, as well as for America's Apollo astronauts. Each country's press had in fact used the word "hero" in its celebrations, which now seemed to hold a distinct importance in the respective narratives that would unfold.

When Neil Armstrong first piloted the *Eagle* to the lunar surface, for instance, the nation learned not only of that achievement but also of the particulars of the hardship that had to be overcome. Less than thirty seconds' worth of fuel remained. Armstrong's pulse raced. And according to biometric data, his heart rate had leaped to 150 beats per minute. The lander was perilously close to an aborted landing, and perhaps even a crash. But then history was made. At 20:17:58 UTC, Armstrong's voice cut through the air. "Houston, Tranquility Base here. The *Eagle* has landed." It was a hero's narrative that could inspire and galvanize. President Barack Obama later dubbed Armstrong "among the greatest of American heroes, not just of his time but of all time." And yet when asked about "the adulation of celebrities and the inflation of heroism," Michael Collins, command module pilot of Apollo 11, actually bristled at the notion. "Heroes abound, and should be revered as such, but don't count astronauts among them. We work very hard; we did our jobs to near perfection, but that was what we had hired on to do. In no way did we meet the criterion of the Congressional Medal of Honor: 'above and beyond the call of duty.' "

That, of course, wouldn't stop the term's use, even as the people themselves usually proved more complex and nuanced than the narrative might allow. In some cases, those hailed as heroes were themselves uneasy with the title, or came to challenge the very systems that elevated them.

Interestingly, Gagarin—the world's first man in space and celebrated hero—became sharply critical of Soviet officials after his friend and colleague Vladimir Komarov was later allowed to fly aboard Soyuz 1, a spacecraft widely regarded by insiders as unfit for launch. True to their concerns, in 1967 Komarov would become the first person known to die during a space mission. As his craft plummeted to Earth, he was reportedly "cursing the people who had put him inside a botched spaceship," according to a damning account in the book *Starman,* written by Jamie Doran and Piers Bizony, who relied on accounts of a KGB officer named Venyamin Ivanovich Russayev and other reporting by Yaroslav Golovanov in the Russian newspaper *Pravda*. Komarov's charred remains were left in an open casket. Gagarin, meanwhile, reportedly became incensed at the Soviet leader, Leonid Brezhnev, and other Soviet officials over the death of his friend. Yet Russayev reportedly warned Gagarin to be careful. "I told him, 'Talk to me first before you do anything. I warn you, be very careful.' " A year later, Gagarin was killed in a plane crash during a routine training flight. Yet he would nonetheless be remembered as a "Hero of the Soviet Union."

Again, that word, "hero," had been invoked, even if the protagonist's perspective had changed.

In ancient Greece, heroes, as direct descendants of both gods and mortals, were effectively demigods wielding supernatural strength and courage. But those bloodlines were not enough to cement their revered status. Adversity, usually through a journey, had to be overcome. The more unlikely the victory, the more compelling the tale. A triumphant return would secure the legend. From Homer's *Odyssey,* where Odysseus emerges from obscurity through bravery and intellect, to more modern times, where Robin Hood, Rocky, and

Luke Skywalker all are central figures with comparable plotlines. The hero must face and overcome adversity.

It would make sense, then, for a nation to use such terminology. Is it any wonder, then, a similar script would follow in China? In that context, the details of Yang's shuddering craft and cracking windows (information that might otherwise be withheld by a government not known for its transparency) started to make sense. It wasn't just the technology. It was the details of the story, and the legend of what it took—along with that longer lens of history—that seemed critical.

Narratives, however, can also have a way of running amok.

· CHAPTER 11 ·

# SCIENCE IN THE HAUNTED 1950S

Every year since 1924, the History of Science Society (HSS) has convened an annual meeting of the nation's brightest scientific minds. Be it war, natural disaster, or pandemic, the conference has gone on virtually uninterrupted in one form or another with the goal of contextualizing scientific knowledge. The point is to both promote science and offer guidance to sidestep the mistakes of the past. In 1990, attendees were encouraged to attend the four-day forum at the Holiday Inn Crowne Plaza in downtown Seattle, where two professors from Brooklyn were scheduled to give a talk. One was a historian of physics named Lawrence Badash. The other was an archivist and historian of science from Caltech named Judith Goodstein. The former had researched top-secret communities that supported the Manhattan Project, a clandestine U.S. initiative where Robert Oppenheimer led a team of physicists to develop the world's first atomic weapons. The latter would go on to chronicle the lives of two Italian mathematicians, Gregorio Ricci and his student Tullio Levi-Civita, whose intellectual contributions in tensor calculus afforded Albert Einstein the mathematical tools he needed in his formulation of general relativity. But on that day, the pair was presenting their combined work on McCarthyism's influence on the scientific community in a session called "Science in the Haunted Fifties."

Badash had been especially vocal on how damaging McCarthy had been, referencing a 1954 White House and Atomic Energy Commission meeting that President Eisenhower had sought to keep secret from Senator Joseph McCarthy. The Republican senator had burnished a reputation for rooting out Communist subverters, often with scant evidence, and the former NATO supreme Allied commander not only had grown weary of it but saw McCarthy as a growing threat to American scientific progress. "We've got to handle this so that all our scientists are not made out to be Reds," Eisenhower once reportedly exclaimed. "That Goddamn McCarthy is just likely to try such a thing."

Until 1950, the junior senator from Wisconsin had been a relatively obscure figure in American politics. But in February, in a show of theatrics during a speech in front of the Ohio County Republican Women's Club in Wheeling, West Virginia, he would produce a piece of paper that he claimed was a list of "205 [State Department employees] that were known to the Secretary of State as being members of the Communist Party and who nevertheless are still working and shaping the policy of the State Department." The move shaped a growing climate of fear, suspicion, and repression that had sweeping effects across American society, just as Eisenhower quietly "loathed McCarthy as much as any human being could possibly loathe another," according to his younger brother Milton Eisenhower, as told by the author John A. Farrell. Dwight's elder brother Arthur had apparently referred to McCarthy as "the most dangerous menace to America." As a presidential candidate in 1952, Eisenhower had briefly campaigned with the Wisconsin senator, even as he worked to discredit him once in office, careful, however, not to attack him directly out of concerns of elevating his status. Still, the fearmongering brand would garner considerable influence.

A 1949 Gallup poll found that roughly 87 percent of respondents believed all Communists should be removed from industries that were thought to be vital to any wartime effort. A year later, a similar poll asked what should be done with Communist Party members in the event of a war with Russia: More than one-fifth responded that they should be put in internment camps. Meanwhile, the effects of McCarthyism began to manifest themselves across the scientific community in research priorities and funding, censorship, and newfound constraints on international collaboration—curiously, just the kinds of conditions much of the science community would complain of generations later under the Trump administration, and the kinds that seemed to benefit Chinese space leadership.

"These types of things only help us," Xue told me over the phone one evening. The former deputy director general of China's National Astronomical Observatories then put it more bluntly. "Our PhDs used to leave for the U.S. and never come back. Now they stay."

Back in 1952, however, Dr. Linus Pauling, then head of Caltech's Department of Chemistry and Chemical Engineering, was feeling the effects of a changing narrative born of a new climate of fear. He said that the U.S. State Department had informed him that it had denied the passport he needed to attend a Royal Society of London conference on proteins because it "would not be in the best interests of the United States." Pauling added that he was informed by a State Department official that the decision had been made "because of suspicion that I was a Communist and because my anti-Communist statements had not been sufficiently strong." Others, meanwhile, were also in the crosshairs. Martin Kamen, an American chemist and son of Russian émigrés whose discoveries in photosynthesis helped unleash a revolution in the carbon dating of fossils, for example, was also accused of leaking atomic bomb secrets following his work on the Manhattan Project. In 1948, he was summoned before the House Un-American Activities Committee, after which time the State Department declined to approve his passport requests. The hunt for Communist spies was not, of course, entirely a fruitless pursuit. By the mid-1940s, the Soviet Union had ratcheted up espionage efforts in a bid to steal American security secrets. In 1949, the CIA director, Roscoe Hillenkoetter, would famously deliver to President Harry Truman a report of "an abnormal radioactive contamination," which he correctly assumed to be an "atomic explosion on the continent of Asia." The Soviets had, in fact, tested their first atomic bomb at the remote Semipalatinsk test site in Kazakhstan, only four years after America's attacks on Hiroshima and Nagasaki. The news proved jarring in the United States, where many officials had assumed Moscow could not progress nearly as quickly as it had, fueling a growing hysteria about subversive Soviet efforts. In fact, the phenomenon had arisen much earlier with the collapse of the Romanov dynasty in Russia, and the subsequent rise of a Marxist Bolshevik party under Vladimir Lenin, along with the growing influence of organized U.S. labor movements. President Woodrow Wilson would then sign into law the Sedition Act of 1918, which essentially extended the Espionage Act from a year earlier and infringed on Americans' rights, among other things, to "utter, print, write or publish any disloyal, profane, scurrilous or abusive language" about the United States during wartime. It would be repealed two years later, but the precedent had been set. In 1950, a controversial piece of legislation known as the McCarran Internal Security Act, enacted over President Harry Truman's veto, essentially required Communist organizations to register with the U.S. Justice Department and called for immediate detention of those suspected of espionage or sabotage.

It was a dark chapter in American history, encouraging powerful lawmakers like the Nevada Democratic senator Pat McCarran, a man long suspected of antisemitism, to also support passage of the Immigration and Nationality Act of 1952. The two laws formed much of the basis for what became a subsequent suppression of ideas and politically motivated prosecutions, often with an anti-immigrant bent—a tack that had already been supported by the earlier Alien Registration Act of 1940 and the Alien Enemies Act of 1798, both of which sought to expand government authority over non-American citizens during times of national emergency or professed threat. Public lists circulated of suspected Communists, including those in the press. Meanwhile, visa restrictions were being tightened as concerns mounted over the hemorrhaging of trade secrets in what was a relatively new atomic era.

As the MIT physicist and science historian David I. Kaiser wrote, the hysteria of the era had placed the science community under enormous scrutiny, with theoretical physicists the "most consistently named whipping-boys of McCarthyism: repeatedly subjected to illegal surveillance by the Federal Bureau of Investigation (FBI), paraded in front of the House Un-American Activities Committee

(HUAC), charged time and again in the media as well as in federal courts with being the 'weakest links' in national security, and widely considered to be more inherently susceptible to Communist propaganda than any other group of scientists or academics." Even Albert Einstein in a 1954 letter to *Reporter* magazine wrote that if he were young again, he "would rather choose to be a plumber or a peddler in the hope to find that modest degree of independence still available under present circumstances."

In the spring of 1950, however, those McCarthy-era policies would set in motion one of America's most costly blunders, helping propel its most significant adversary since the Soviets into space. Former secretary of the navy Dan Kimball called it "the stupidest thing this country ever did."

The year was 1955. And a man named Tsien Hsue-shen would soon find himself at a crossroads. Born in the heart of China's eastern Zhejiang province, Tsien had traveled to the United States at the age of twenty-six, earning engineering scholarships at MIT and Caltech. His father taught philosophy and ethics before marrying into a family of politically connected silk merchants in Hangzhou, where pagoda-topped hills surrounded the city's West Lake, a freshwater body long the focus of poets and painters. While there, his son studied physics, chemistry, and mathematics, as well as engineering in neighboring Shanghai, at Jiaotong University, where Western influence had over the generations become more prevalent. Loosely translated as "On the Sea," Shanghai had been made a colonial city after the so-called Opium Wars, and had thus emerged as a key trading port, with a more market-driven economy, as well as more noticeable disparities between rich and poor. Foreign-run cafés and theaters would earn it the moniker the Paris of the East, which might have influenced Tsien's own Western interests. Once in the United States, he had begun working under the famed Hungarian mathematician and aerospace engineer Theodore von Kármán, who was by then the head of Caltech's Guggenheim Aeronautical Lab in Pasadena. Von Kármán had taken to the brilliant young Tsien and asked him to join his core group of high-altitude rocket enthusiasts, exploring and making advances in solid- and liquid-fuel propulsion, which at times involved employing jet engines mounted onto prop planes. Tests had been carried out in the nearby Arroyo Seco canyon, a remote stretch north of Pasadena acquired for the team in 1936 by the U.S. Army Air Corps after Allied intelligence had learned of the German V-2 rocket program. And Tsien, along with the aeronautical engineer Frank Malina, whom NASA credits as one of the founders of modern rocketry, as well as the engineers Jack Parsons and Ed Forman, would play a foundational role in the formation of NASA's Jet Propulsion Laboratory, the soon-to-be-famous research hub that would go on to develop America's first satellite, probes to Venus and Mercury, the Mars rover, and humanity's first craft to depart the solar system.

By 1950, however, it was still early days, and the so-called Red Scare seemed to grow scarier, especially for Tsien. FBI agents were now actively scrutinizing his old political ties. Just a year earlier, the same year China's Communists under Mao Zedong had driven the country's Nationalists under Chiang Kai-shek into Taiwan exile, Tsien had sought U.S. citizenship. But after revelations about his attendance at graduate school parties years earlier, organized by members of Pasadena's local Communist Party, his American loyalties were now in question. In June, federal agents visited his home. His security clearance would be revoked. Jailed, and placed under house arrest for some five years, he was unable to visit his ailing father, despite pleas of innocence. Factors like his wife being the daughter of a Nationalist opposition leader, along with his nearly two-decade residence in the United States, as well as letters from those like von Kármán, appealing on Tsien's behalf, would not help.

The Chinese-born rocket scientist would be deported back to China in 1955. And yet he wouldn't arrive in time to see his father, who had died while Tsien waited under house arrest. Embittered and perhaps now especially vulnerable to Communist Party influence, it should not be surprising that his return coincided with the beginnings of China's space program. Heralded as a kind of Chinese Wernher von Braun, he would eventually come to be known as the father of the country's ballistic missile and space program, contributing to China's first satellite, the Dong Fang Hong 1.

It was a compelling story: A prominent Chinese-born rocket scientist contributing to the U.S. military's rocket program was expelled and then went on to "revolutionize . . . the whole of missile science in China," according to the later testimony of Ernest Kuh, who served as dean of engineering at Berkeley. But still, it was a story that was, at the time, hardly known to the broader public. Badash and Goodstein had resurfaced it during their presentation nearly half a century later at that HSS conference in Seattle, with the aim of highlighting how so-called witch hunts of the past could not only hinder U.S. technological advancement but also strengthen its rivals.

Luckily, however, in attendance at that forum was a senior editor with HarperCollins Publishers named Susan Rabiner, who seemed to immediately recognize the story's value. Searching for authors to unearth Tsien's account, she offered the project to a twenty-one-year-old graduate student named Iris Chang, who was the daughter of two university professors who had moved to Princeton, New Jersey, from China, by way of Taiwan. Chang's subsequent reporting would eventually become the most comprehensive accounting of Tsien's life, and the nature of his expulsion. In interviews with family members, colleagues, and friends, as well as in reviewing government and university documents, Chang pieced together Tsien's evolution, and the bitter resentment he eventually manifested against the United States as a result of his deportation. In a rare interview, Tsien's son, who worked for a Taiwanese-owned computer company in Fremont, California, explained how "it was like having someone as a guest in your house and later kicking him out.

"If my father had committed a crime in this country, then my father would have nothing to say. But my father devoted twenty years of his life to service in the United States and contributed to much of this nation's technology, only to be repaid by being driven out of the country," he told Chang.

"He was no more a Communist than I was," the former navy secretary Dan Kimball later explained. "And we forced him to go." In the interim, Tsien—as Chang writes—found himself drawn closer into the Communist orbit. "How stark the contrast between the young Tsien and the old," she wrote. "The young Tsien dreamed of a world of peace and equality. The older Tsien lived in a world governed by regimented hierarchy and helped manufacture the weapons of world destruction. The young Tsien was both Chinese and American, at heart a citizen of two countries. The older Tsien felt alienated by both."

In 2009, he died in Beijing, at the age of ninety-seven, after having built the Dongfeng missile program as well as the country's Silkworm anti-ship cruise missiles, which were sold to the Iraqi military and ironically used against the USS *Missouri,* an American battleship, during the first Gulf War. More broadly, Tsien's contributions to China's missile capabilities seemed to lay the bedrock not only for the country's growing ambitions in space but also for a strategic shift that would over time contribute to a recalibration of U.S. defense priorities. Only a decade later, America's military would grow to include a sixth branch, shifting its focus to prepare for the very adversary that Tsien had propelled into space. With the stroke of a pen, President Trump would create the U.S. Space Force, later pushing for the agency to receive a budget nearly double that of NASA. The militarization of space—something signatories of the Outer Space Treaty had sought to prevent—was fast becoming an inescapable reality.

Space now expressly seemed like a war fighter's domain.

Yet the newfound branch would soon face challenges of its own.

· CHAPTER 12 ·

# THE PENTAGON'S NEW RIDE

At 2:17 a.m., a new kind of rocket was about to blast off. Of the Space Coast's hundreds of yearly launches, this one was unique. Bathed in floodlights at Cape Canaveral Space Force Station in Florida, the Vulcan Centaur, a product of United Launch Alliance (ULA)—a joint venture between Boeing and Lockheed Martin—was just sixty seconds from taking flight. Space Force brigadier general Kristin L. Panzenhagen called it "a groundbreaking day," marking the "transition from foreign engines."

Until now, outside an expanding fleet of Falcon 9s and Falcon Heavies, the Pentagon had generally looked to just two rockets to deliver its payloads into orbit: One was Boeing's Delta IV. The other was Lockheed's Atlas V. Both were nearing retirement. But it was the Atlas V that posed the bigger political problem. Its engine—the RD-180—had been built by a Russia-based company, which was subject to sanctions after Moscow invaded Ukraine. The U.S. Congress had instructed the Pentagon to put an end to the use of Russian launch vehicles and technology. But the engine couldn't just be swapped out. RD-180 was integrated into the overall structure of the Atlas V. Removing it would compel a full design overhaul. Replacing it just wasn't an option.

"That's not how rockets work," United Launch Alliance's CEO, Tory Bruno, told Mike Gruss of *SpaceNews*. "The engine would be designed for the rocket, or the rocket would be designed for the engine. There really isn't anything available for a drop-in. That's just not how it's done." That then compelled lawmakers to ban the future use of the engine, while affording a compromise that allowed for waivers when it came to national security missions.

The idea, however, was to eventually phase the engine out altogether. No matter how reliable it had been, the days of the RD-180—and by association the Atlas V rocket—were numbered. Still, before closing the chapter on this storied piece of engineering, it's worth looking back at the remarkable history behind it, especially where it was built. Situated just eleven miles northwest of the Russian capital, the area had been reached by German soldiers during World War II, before their advance stalled in late 1941. But what would happen there in coming years would work to change the very standards of rocketry.

By the war's end, that Russian city of Khimki had become home to a guided missile plant known as Plant No. 456. And according to a declassified 1957 CIA report, it soon became filled with "numerous German scientists and technical personnel who had been involved in the V-2 (rocket) program." Inside, workers pulled two shifts per day, six days per week, and worked on building the seven-meter-long Fau-2 rocket—essentially a Soviet version of the German V-2 and a precursor to modern space rockets and intercontinental missiles.

"At irregular intervals, a very loud, ear-shattering noise emanated from the airfield. This noise lasted for about one minute," the report said. "They were testing some sort of engine."

Outside the facility, and at each entrance into the plant, revolver-wielding guards in gray uniforms could be seen as workers deposited their entry passes and retrieved them upon leaving. The assembly shop, meanwhile, consisted of a single-story brick building with a sheet-metal roof and was where "the guided missiles were assembled." That same year, 1946, the Soviet design bureau OKB-456 was formed, responsible for developing more than 120 first-stage engines for the Soviet intercontinental ballistic missiles program. Founded by the rocket engineer and Soviet space program founder Sergei Korolev, it was located in Khimki. And the work done there would in 1957 develop the world's first ICBM—the R-7 rocket—capable of reaching orbit and delivering a nuclear payload that could hit the continental United States. Over time, like many Soviet-era design bureaus, OKB-456 would be absorbed into larger organizations as the space industry matured and restructured. Eventually, that work was integrated into a major Russian aerospace company, NPO Energomash, still active today and recognized as a leading Russia enterprise in the development of liquid-propelled rocket engines, including the RD-180. In other words, the descendants of the very design teams that once put Washington, D.C., in Soviet crosshairs had over time started building rocket engines for the Pentagon itself. With its dual-nozzle engine and dual-combustion chamber, which benefited from a proprietary metal alloy that prevented cracks, the system was considered highly reliable. The Russians just knew how to build good rocket engines, with the Atlas V having a nearly flawless launch record. In fact, in 2019, after SpaceX successfully tested its Raptor engines, Elon Musk tweeted that his company had only then finally surpassed its Russian counterparts, having in many ways embraced Moscow's more iterative approach.

"It's amazing that the RD-170 & RD-180 engines held the record for so many decades. Excellent engineering," Musk wrote. Part of it, also, might have been a product of a Soviet design culture with fewer guardrails. The engine's design initially drew skeptics, given its use of high-pressure oxygen from the turbine exhaust created significantly higher pressure within the engine's combustion chamber than had been permitted under the more risk-averse U.S. standards. But the design also required considerably less energy to kick-start the combustion process. It was a risk-reward scenario that could result in a far more efficient system. Russian manufacturing was thought to allow higher flexibility (and risk) in the design process. Through hundreds of iterations, along with the invention of a new kind of high-temperature-resistant stainless steel, the RD-180 was born. By 1997, during a time of warmer relations, Lockheed Martin had struck a deal with NPO Energomash to purchase them for its Atlas rockets.

Now newer American-made engines were meant to take up the slack. Blue Origin's founder, Jeff Bezos, had poured millions into a testing facility in West Texas for just that purpose. And yet during a static fire test in the summer of 2023, the company's BE-4 engine detonated after only ten seconds, destroying the surrounding infrastructure while threatening further delays of the Vulcan launch. ULA had suffered its own setback in March in an explosion during structural testing. But in January 2024, the new 202-foot-tall rocket was ready—the culmination of nearly a decade of development meant to wean America's defense sector off Russian technology.

Of course, it also served another purpose: a much-awaited challenge to SpaceX's market dominance. By then, the SpaceX Falcon 9 rocket had all but controlled the launch industry, which meant the military had precious few alternatives. Eventually, under the Trump administration, SpaceX would cement its spot as the Pentagon's leading launch provider with a $5.9 billion deal. But only a few years earlier, the Department of Defense had selected ULA to carry out nearly two-thirds of all national security missions. Production delays, however, had contributed to a backlog of some seventy missions. A year later, in January 2024, the Pentagon sought to take the reins entirely, and asked to oversee SpaceX's Starship for its own "sensitive and potentially dangerous missions," yet another sign of the deepening federal reliance on the Musk-led company. Vulcan Centaur had been meant to spread out that dependency.

But it was also important for another reason. Riding inside its cone was *Peregrine,* a rival Moon lander built by the Pittsburgh-based company Astrobotic, and a direct challenger to *Odie,* the sleek, hexagonal craft from Intuitive Machines that once seemed destined to be first. With Vulcan now airborne, the Moon race suddenly seemed alive again. And yet inside mission control, Astrobotic's team was now scrambling.

Something had gone terribly wrong.

· CHAPTER 13 ·

# MOON RUSH

Back in Houston, the Intuitive Machines team was dealing with its own set of problems. Rob was back on headsets. And from the looks of it, he seemed to have all but disappeared into his own distant world of hisses and crackles. Changes in frequency or tone, small hissing or sputtering sounds, or a thumping or rumbling could all indicate bigger problems. "Machines talk to us," shuttle astronaut Danny Olivas explained, "if we are willing to listen."

Rob, for his part, was tuned in. This time, he eased the throttle down, which made the distinct popping noise more apparent. By adjusting thrust, he had hoped to discern just how the engine responded at different power levels, effectively testing the machine's efficiency and stability. If he could isolate the noise, he could start to work the problem.

Still, there were limits. Earth-based tests were never quite like the real thing in space. Microgravity affects how propellant flows, how heat is transferred, how fuel and oxidizer mix, and even the shape of the flame. It was trial and error, and frankly guesswork. Minor problems on Earth could be mission ending up there. Plus, the gravity would change again when *Odie* approached the Moon. Unlike an aircraft, which descends onto a runway by reducing power and using its flaps to create drag, landing a lunar craft effectively requires it to steer itself into the ground. To land, *Odie* would have to thrust downward at speeds of roughly one meter per second, throttling down to no more than 20 percent of max thrust. Too much, and it would never reach the surface. Too little, and it would crash. The team had gone through scores of injector configurations. But it still wasn't quite right.

"All right," Rob said, sliding his chair back and tearing off headsets. "That's enough." The tube whined and squeaked as the remaining fuel cut off. A few remaining put-puts gave way to a thin trail of black smoke.

"We'll do it again," he said, now sitting back in his chair. He seemed tired. They all did, having been at the range since morning. Failure, however, was not always so obvious. In those early days, the only way one could know for sure if the system was working as intended, he explained, was by looking at the raw data. The team would test-fire all day, but it was at night when the real answers came.

Downloaded onto a thumb drive, it would "be burning a hole in my pants," he recalled, having crammed the data into his pocket before heading home. What was the fuel consumption? How did the engine perform? Where were the stress points? Optimizing a rocket engine requires patience. Then again, this wasn't NASA. Too many tweaks and refinements cost time and money that chewed into profit margins.

Still, it'd have to work.

At first, the combustion was happening too far from the injector, which meant the cryogenics that cool the rocket face were causing too much condensation. At -300°F, ice buildup at that range was almost inevitable. SpaceX had experienced a very public crash landing in 2016 for similar reasons. Rob's team adjusted. But this time, they overcorrected. The cocktail was now forcing combustion too much the other way. Methane and oxygen were mixing inside the injector, causing a series of micro-detonations inside the engine itself. The result was a violent shudder and shake. Not the kind of thing one would want while landing on the Moon. Ultimately, good enough would have to do. They would have to launch. Engineers like Rob, who preferred to continuously test and improve, had to get used to the idea of "good enough." And yet, for Intuitive Machines, both the timetable and the landing zone were about to change.

As Rob's team toiled with propellant ratios and injector configurations, Steve's phone was ringing. On the other end of the line was Bill Nelson, the former Florida senator turned NASA administrator, who had apparently grown even more alarmed by Chinese ambitions. Beijing's Chang'e 7 mission—a robotic expedition to the lunar south pole—had targeted one of the very same landing sites proposed for NASA's Artemis 3 mission. A cone-shaped impact crater at the lunar south pole, the rim of Shackleton crater was coveted for at least two reasons: (1) Parts of it were shaded in permanent darkness and extremely low temperatures that made for ideal conditions for those all-important water-ice deposits; and (2) given the Moon tilts on an axis of only 1.5°, compared with Earth's more dramatic axial tilt of roughly 23.44°, other parts were bathed in near-continuous sunlight—critical for drawing solar power and allowing astronauts to work in the light. But it wasn't just Shackleton. Chinese plans also pointed to overlapping sites at similar craters, including Haworth, Nobile, Amundsen, and Malapert.

Nelson was sounding the alarm. In fact, during a House Science, Space, and Technology Committee hearing in 2023, he held up an image of future Artemis landing sites to show how almost walkably close they were to their Chinese counterparts. "This is where we're going," he said for effect. "This is where China is going."

Water, of course, was the recurring focus. To facilitate at scale, NASA had devised plans for lunar bottling plants, and even a five-kilometer oxygen pipeline across the Moon's south pole, to be built by robots and 3D printers—using locally mined materials—for the purpose of delivering extracted oxygen directly to human habitats and liquefaction plants. Other resources, including trace amounts of helium-3, only added to the stakes, particularly after China reported its discovery in 2020 within a lunar sample. And yet as competition intensified, the two agencies were restricted from engaging in direct dialogue. In the United States, the Wolf Amendment, named after Representative Frank Wolf (R-Va.), prohibits NASA from cooperating with the Chinese government or affiliated entities as a means of safeguarding U.S. interests and intellectual property, but also as a punitive measure meant to pressure China over its human rights record. Wolf, then chairman of the House Appropriations subcommittee that funds NASA, outlined his concerns in a 2013 letter to the NASA administrator at the time, Charles Bolden.

"In light of these realities," he wrote, "I have supported efforts to limit new collaboration with China until we see improvements in its human rights record, as well as a reduction in its well documented cyberattacks and espionage efforts against the U.S."

At least partly a result of that inability to converse directly, perhaps, the rhetoric was ratcheting ever higher.

Nelson accused his Chinese counterparts of laying claim to the territory, prompting a stern response from Beijing, with its Foreign Affairs spokesman Zhao Lijian claiming Nelson was lying "through his teeth." Tensions escalated during a 2023 U.S. embassy briefing in London, when Brigadier General Jesse Morehouse of U.S. Space Command pointed to Russian moves and China's advancing anti-satellite tech, warning that Washington had "no choice" but to prepare for orbital warfare; especially given that Space Force officials were now tracking new "dogfighting" tactics by Chinese satellites.

"The United States of America is ready to fight tonight in space if we have to."

In Beijing, meanwhile, the rapid development of hypersonic missiles, anti-satellite weaponry, and military space planes seemed to only add to the commentary, just as the topic of the Moon kept emerging. A journal article, authored in part by Zhang He, the Chang'e 4 lunar mission commander, called it a "springboard" for the rest of space. In that article, he referred to the Moon as "a strategic point for all space powers to compete."

NASA now needed the Houston team to hustle.

"Can you get there by March?" Bill Nelson asked over the phone, according to Steve. "We want two missions before the Chinese show up." Speed was key. China had thrice completed lunar landings since 2013 while also becoming the first nation to send a rover to its far side. In 2020, the Chang'e 5 mission launched from Wenchang Space Launch Center on Hainan Island aboard a Long March 5 rocket, and had landed on a part of the Moon known as Oceanus Procellarum, or Ocean of Storms. Two of its modules collected lunar soil samples and returned them to Earth—the first time that had been done in more than forty years. The Chang'e 6 mission would do it again, only this time on the Moon's far side, returning the first lunar far-side samples following a fifty-three-day mission. Meanwhile, China's main space agency was about to put out the call for Chinese developers to begin developing the core technology for equipment capable of using energy collected from the Sun to melt lunar soil and mold the resulting goop into bricks. If successful, those bricks could then serve as the foundation of what Chinese leadership hoped would leapfrog Artemis and complete the first permanent human outpost on the Moon, operational by 2035, with plans for a launchpad that could service crewed missions across the solar system by 2050. Many in Washington, meanwhile, had become increasingly concerned that while China was advancing a sustainable long-term strategy, Artemis was more focused on the photo op of a flags-and-footprints strategy that had ultimately stalled her predecessor, Apollo. The proposed Artemis Base Camp, for instance, had comparably little in the way of public details.

The priority then seemed to focus on an already over-budget and delayed Moon-orbiting Gateway space station, to be built in stages, and which NASA considered "central" to the mission, even though none of the Apollo missions had ever employed such an architecture. Plus, Gateway not only had no reliable launch date, but it was very clearly on the Trump administration's chopping block.

"Why would you want to send a crew to an intermediate point in space, pick up a lander there and go down?" asked Buzz Aldrin, who later referred to the Gateway project as "absurd." Then, in July 2024, the U.S. Government Accountability Office got specific, identifying budget overruns and "several significant challenges," much of which centered on Gateway systems, including the lack of an "overall mass management plan."

The entire system seemed destined to be scrapped. There were, however, caveats.

"Gateway is useful when, but not before, we are manufacturing propellant on the Moon and shipping it up to a depot in lunar orbit," former NASA administrator Mike Griffin noted. If the Moon were to be used as a kind of gas station and launchpad, from which a spacecraft might fill up and go explore the broader solar system, perhaps Gateway could serve a greater function. Based on current trends, however, any such "gas station" was more likely to fall under the domain of the commercial sector, not NASA. Others like David Brin, a longtime consultant for the agency and science fiction writer, fell somewhere in the middle, advocating for development of near-Earth asteroids that have "way more water," but also recognizing the possible value of Gateway as a potential way station for exploration, search and rescue, and research. Still, if Mars was the bigger goal, devoting years to building an orbiting lunar station was at the very least debatable. The concern, meanwhile, was that while Gateway orbited the Moon, Chinese efforts would be fast developing a more robust presence on the lunar surface itself, with superior capabilities to harvest resources and manufacture infrastructure in a way that could far outpace Artemis.

There just seemed to be more value in actually being on the Moon, as opposed to orbiting it.

· CHAPTER 14 ·

# RED STAR AND CLOVER

Eight thousand miles west of Houston, the light of daybreak spilled over the banks of the Yangtze River, which cuts through the sprawling metropolis of Wuhan. There, a clatter of bicycles mixed with the steaming smells of thousands of breakfast stalls, as commuters edged their way past skyscrapers and ancient temples. It was late April 2023. Scientists and engineers from across China's space sector were convening for the Chinese Society of Astronautics and the China Space Foundation's annual two-day conference, which also commemorated the launch of China's first satellite. Over the years, Wuhan had emerged as both a major industrial center and a space research hub. It was an identity that somehow persisted even after the city gained global attention as the epicenter of the first known cases of the novel coronavirus, SARS-CoV-2. A year earlier, in 2022, the city had offered the equivalent of up to nearly $8 million each to rocket and satellite companies, with the aim of becoming China's "valley of satellites"—a nearly $16 billion push that supported more than three hundred commercial aerospace companies across the region. But at that conference in the spring of 2023, where grandiose visions of Mars and Jupiter abounded and talk of intelligent life, quantum frontiers, and the mysteries of black holes, gravity, and relativity electrified the room, it was a man named Ding Lieyun who rose to speak with something considerably more tangible.

Local to the region, having studied at the nearby Wuhan Polytechnic University, Ding was an engineer by trade, having played key roles in devising digital management systems for China's rail lines.

But after he served as director of Huazhong University's National Laboratory of Defense Technology, his focus seemed to narrow on space. By the summer of 2023, as chief scientist of the National Centre of Technology Innovation for Digital Construction, he was now focused on "extraterrestrial construction."

It was obviously "still at a very early stage," but "building habitation beyond the earth [was] essential not only for all humanity's quest for space exploration, but also for China's strategic needs as a space power," he said. Two years earlier, his nation had launched the core module of the Tiangong space station, China's orbiting laboratory, a fraction of the size of the ISS, but designed for expansion, with plans for an eventual 180-ton, six-module system, replete with inflatable habitats and six docking ports, to be occupied by a rotating crew of three. Now he was turning to some of China's brightest space engineers, not just for conjecture and theory, but for actual design ideas to help refine two working systems that could sustain human life on the Moon. The blueprint for a permanent lunar habitat was taking shape. And Ding Lieyun was at the center of it. But he wasn't alone. As the discussion got under way, Yu Dengyun, China's deputy chief designer for its lunar exploration program, laid out the two options: One was called Clover. The other, Red Star.

Clover looked like a three-leafed clover, beginning with inflatable habitats. Once inflated, rover-like robots (dubbed Super Masons) would gather lunar regolith to make bricks that would surround the inflatables to harden them, with the first basic structure planned for as early as 2028, along with eventual solar and nuclear energy powering the base as it expanded.

Red Star, by comparison, would employ a more elemental approach, making use of lunar craters with four cabins to form the four points of a starlike structure. Much like humanity's earliest shelters on Earth, the design could be adapted to make use of the Moon's vast network of lava tubes, transforming them into massive subterranean habitats. First discovered in 2009 by the Lunar Reconnaissance Orbiter and Japan's Kaguya spacecraft, lunar caves were thought of as potential entrances to tunnels, formed by ancient basaltic lava flows billions of years ago during a time of volcanic activity across the Moon. When the lava eventually drained away and hardened, hollow tubes had been left behind, which engineers in Wuhan were reviewing as the potential foundations for base camp. If it worked, not only would those tubes presumably require less construction to create human habitats, but they also afforded a degree of climate control and protection from radiation and meteorites. These were not new ideas. Research papers dating back to at least 1962, if not earlier, showed comparable designs. The difference now was that the Chinese were operationalizing them.

Yu, whose team had devised earlier plans to use those Super Mason robots to develop a more egg-shaped structure, using Lego-like 3D-printed bricks, was now discussing ways to simulate the lunar environment on Earth to make it happen. The challenges were immense; not just in getting there, but in operating in such punishing environments. Temperatures on the lunar south pole could range from the peak of an Arizona summer to far colder than anything experienced on Earth (-392°F). This wasn't just a design problem, but rather a battle against the kind of extremes no human structure had ever endured. "Such a difference is larger than we had expected and adds to the difficulty of in-situ construction." Measurements taken at a lunar impact crater located close to the lunar north pole were found to be even colder, at -250°C (-410°F), constituting among the coldest temperatures measured anywhere in the solar system. It was colder than Pluto. And yet NASA's Lunar Reconnaissance Orbiter had discovered that caves at the lunar surface worked to buffer the Moon's extreme day-night temperature swings, with the climate hovering around a consistent (and rather enjoyable) 63°F.

"Humans evolved living in caves, and to caves we might return when we live on the Moon," explained David Paige, who oversaw the Diviner Lunar Radiometer Experiment that conducted the measurements used in the study.

These insights, however, were fast becoming less theoretical. Just a few months later, some of the same scientists and engineers who had met in Wuhan had gathered again, this time in Huzhou City. There, at a follow-up conference, the precise ways in which humans could live on the Moon seemed to be taking a more practical turn. In that presentation, an animation showed an orbiting craft firing a drilling device at the lunar surface. The device penetrates the regolith and bores into the tubular network beneath the surface. Robots then reinforce the opening of the blast site, preparing it for the landing of the core module inside the resulting hole. From there, subterranean structures are built and expanded up under the (hopeful) protection of the surrounding edifice. Meanwhile, a growing lunar constellation of satellites would serve as basic relay communications and navigation, with the potential to grow. Still, much of the talk centered on speed, and the urgency of the moment.

"Now is the critical time for space infrastructure to expand to the Earth-moon system," explained Yang Mengfei, a top space policy adviser to Beijing. In a March 2023 statement, representing CASC, China's main space contractor, he described it as a chance that will "never come again."

Yang knew this. But so did many of those at NASA, which had just asked Steve Altemus back in Houston to pick up the pace, advancing his timetable by more than a year. The revised plan had also moved the landing zone to prioritize the lunar south pole, with a subsequent mission there to drill for water ice. National and commercial competitors were now making their moves. Among the latter was the *Peregrine* lander, which was—at the moment—hurtling through space and on its way to the Moon.

· CHAPTER 15 ·

# PEREGRINE

Alone in void and destined for the lunar surface, Astrobotic's *Peregrine* craft had by now separated from the Vulcan upper stage and slipped into a highly elliptical translunar trajectory. The plan was to circle at an altitude of about a hundred kilometers above the Moon before descending to Sinus Viscositatis (the Bay of Stickiness), named for hardened nearby lava flows. From there, the craft would conduct a series of NASA and commercial experiments, which included a rover from Mexico, a radiation detector from Germany, and three NASA instruments that could be employed in the search for water. But more importantly, this four-legged, foil-encased boxy craft sought to make history. And so far, so good. The Vulcan launch had been virtually flawless, as was separation from the rocket's upper stage.

"*Peregrine* powered on, acquired a signal with Earth and is now moving through space on its way to the moon," John Thornton, chief executive of Astrobotic, said in a statement after the launch. Still, there was trepidation in his words.

"We know we're headed into a gauntlet here," Thornton said in an earlier interview. "We know we're headed into very difficult territory."

The video feed had shown a standing ovation inside the United Launch Alliance control room shortly after the rocket thundered skyward and successfully separated. Applause roared over the broadcast. But afterward, and inside Astrobotic control, it seemed more serious. *Peregrine* still needed to correct its trajectory. Its mounted solar cells, which recharged the craft's lithium-ion batteries, had to be facing the Sun. But still they hadn't turned. Mission control was now scrambling to stabilize the vehicle, and was forced to continue using the attitude control system thrusters to keep the ship from tumbling. "Unfortunately, an anomaly then occurred, which prevented Astrobotic from achieving a stable sun-pointing orientation," the company said in a statement shortly after launch. "The team is responding in real time as the situation unfolds and will be providing updates as more data is obtained and analyzed." Hope wasn't gone quite yet. But the team was now clearly fighting for their craft's survival.

About an hour later, Astrobotic put out another statement. "We continue to gather data and report our best assessment of what we see," it said. "The team believes that the likely cause of the unstable sun-pointing is a propulsion anomaly that, if proven true, threatens the ability of the spacecraft to soft land on the Moon."

Time was running out. A subsequent company statement would point to a faulty helium pressure control valve, most likely a result of "vibration-initiated relaxation between threaded components internal to the valve." Still, they managed to get the craft pointed back toward the Sun to start recharging its batteries. But soon it would become clear that it was too late. Leadership had to make the call. In the company's fourth update of the day, on January 8, 2024, Astrobotic put out the following statement:

> Unfortunately, it appears the failure within the propulsion system is causing a critical loss of propellant. The team is working to try and stabilize this loss, but given the situation, we have prioritized maximizing the science and data we can capture. We are currently assessing what alternative mission profiles may be feasible at this time.

The reality was now starkly clear. Another private lunar lander had failed. More significantly, the commercial viability of the Commercial Lunar Payload Services (CLPS) program seemed to be in question. So far, no commercial company had proven able to reach the Moon. And concerns were growing as to whether this was indeed the right tack. Perhaps those old Apollo days had made it look easier than it was. History, however, belies the reality of lunar landings. More than eighty craft, as part of seven space programs and one private company, had crashed into the lunar surface, a number that does not include those that never even reached it. In just the past few years, Russia's Luna 25, ispace's Hakuto-R (and later, its subsequent mission, Resilience), one of India's *Vikram* Moon landers, as well as SpaceIL's Beresheet had all added to the debris strewn across the lunar surface.

Meanwhile, Steve Altemus had been watching. He knew the team at Astrobotic. More than that, he knew what it was like to lose a vehicle: years of work vanishing in seconds. But he also knew the stakes involved, not just for the CLPS program and America's growing competition with China, but also as it related to the perils of human spaceflight that would one day again take shape beyond Earth's orbits. *That* was a far riskier endeavor. Robotic landers could tolerate all manner of iteration. But when humans were on board, the math changed.

Steve understood this in a way few others could. Unlike *Peregrine,* the question of human lives in space seemed to be the nagging uncertainty in this new commercial era. SpaceX's Dragon capsule had successfully ferried astronauts to the ISS, just as space tourists ventured into orbit. But the Moon was altogether different. Eventually, people would descend in these crafts and onto one of the most difficult landing zones the solar system had to offer.

Failure was a very real possibility.

· CHAPTER 16 ·

# FINAL DESCENT—*COLUMBIA*

Two thousand three was a seminal year for NASA. The International Space Station was growing. Six shuttle missions had been planned. And the oldest orbiter in the shuttle fleet was on its way home, marking a close to its first mission of the year. A full three years in the making, the vehicle had been plagued by delays, having to contend with servicing missions for both the ISS and the Hubble Space Telescope. Then again, for *Columbia,* this was nothing new. The orbiter, which had been overhauled three times between 1984 and 1994, had also experienced in-flight anomalies in virtually every one of its missions. And yet, according to a *Washington Post* review of thousands of NASA documents, between 1996 and 1999 the vehicle also experienced "at least five 'escapes'—a NASA term for a mission that flew with a problem that only 'luck or providence' prevented from causing serious damage." The veracity of that reporting could not be independently verified. But an unrelated accident investigation board report years later would describe a "broken safety culture," which had not learned "the bitter lessons" of the 1986 *Challenger* disaster that killed all seven astronauts on board. *Columbia,* being the oldest, was of particular concern.

A visit to Houston, however, had a way of always offering fresh insights. By now, I was making regular trips. And on this particular jaunt, one of *Columbia*'s former pilots had invited me to lunch, a gentleman of a man named Brewster Shaw. A veteran of three space shuttle missions, Brewster had flown *Columbia* in 1983, served as commander of *Atlantis* in 1985, and again as commander of *Colum-*

With the retired NASA astronaut, U.S. Air Force colonel, and former Boeing executive Brewster Shaw for lunch at an old NASA haunt called Frenchies in Houston, Texas, where the walls are festooned with photographs of NASA luminaries. Brewster became indispensable in the writing of this book, both in opening figurative doors and in serving as a kind of gut check on the reporting. Shaw was inducted into the U.S. Astronaut Hall of Fame on May 6, 2006, as a veteran of three space shuttle missions.

*bia* in 1989, and yet exhibited virtually none of the bravado one might expect of a person with a résumé like that. Born in Michigan in 1945, then studying engineering at the University of Wisconsin–Madison, he would join the air force in 1969 and find his way to test pilot school at Edwards Air Force Base in California before eventually piloting *Columbia* into space.

When we met, however, he seemed almost deferential, using phrases like "you probably know better than me." When it came to space, there was almost nowhere that was true. Yet it was a characteristic that I had come across time and again. Despite space cowboy personas conveyed in popular culture, in person there was a distinct humility to many of those who actually worked in space. Perhaps it was born of the self-scrutiny demanded of those at NASA, imbued as it was with a kind of science-based questioning, along with a decidedly more intimate understanding of the scope of the dangers faced. Space missions were inherently fraught. Given that, and the thin threads that separated success from catastrophe, perhaps it was quiet competence that was valued more than braggadocious behavior. Brewster, embodied that sensibility, with an almost disarming modesty.

In truth, without him, much of this book would never have been written. Upon introduction, he had proven a key interlocutor in those early days. And ahead of our meeting, he had now picked what looked like an inauspicious locale, tucked away in a strip mall near NASA Parkway, not far from the Johnson Space Center. Festooned with signed photos of astronauts and other NASA luminaries, Frenchies is an Italian restaurant and popular meeting spot for those in the industry, run by a man named Francesco, who had apparently left Italy for Texas when his sister married a NASA engineer. There, muffulettas were served to the lilting tunes of timeless Italian love songs.

"Let's go back to 1983," Brewster said, as we began to eat.

It was the year of a ten-day shuttle mission that began in late November, meant to deliver the first Spacelab laboratory module. Brewster was chosen to pilot the vehicle, which had just completed its 166th orbit and was now returning for a landing at Edwards Air Force Base. Everything had generally gone smoothly, until, that is, reentry procedures began. Initially, two of the four general-purpose computers (GPCs) on board failed. "One of them recovered," explained Brewster. "The other computer did not." Then there was a problem with the gimbal bearing. "It was banging," Brewster said, so much so that Commander John Young, who had been trying to get a bit of rest before deorbiting, poked his head in.

"What's all this noise?" he asked. "You guys are making this banging noise."

"We told him, 'Well, no, that wasn't us,' " Brewster recalled. "That was an IMU failing. His eyes got bigger after two computer failures and now an IMU failure." IMU is short for inertial measurement unit, which is an electronic box that consists of gyroscopes, accelerometers, and other devices for measuring things like the shuttle's orientation and velocity.

"The first thing you do when a GPC fails is you get out the emergency checklist and you go through the one GPC failure procedure, and then another one fails," Brewster explained over his salad. "We have two GPC failure procedures, and so I'm reading the checklist to John."

What they didn't realize was that a series of fatigue cracks in the hydrazine tanks, used to power the craft's hydraulic system, were now leaking fuel. As the craft deorbited, and more and more of an oxygen-rich atmosphere enveloped the craft, it ignited. A fire had broken out in the back of the shuttle. Descending into lower atmospheres only fed the blaze, scorching valves in the power units and damaging the adjacent wiring. Onlookers reported seeing smoke and yellow flames in the back of the craft, although NASA officials later said they were unrelated to the fire and that "the crew was never in any danger."

In the end, the vehicle would make it home safely. But the mission had highlighted how even small complications could escalate into mission-jeopardizing problems, marking the importance of rigorous safety protocols that critics would later argue had become too lax. Two decades later, those kinds of issues would come to a head during *Columbia*'s twenty-eighth flight in January 2003.

During that mission, there were problems almost from the outset. Just eighty-two seconds into launch, a piece of foam insulation detached from the shuttle's external tank and struck *Columbia*'s left wing, which engineers would later discover while examining footage of the launch. The flight director, J. Steve Stich, emailed Commander Rick Husband and the pilot, Willie McCool, about what had happened. Because foam strikes had occurred during other flights, he reassured them the situation was not mission critical.

"The issue is not even worth mentioning other than wanting to make sure that you are not surprised by it in a question from a reporter," he wrote. "Experts have reviewed the high-speed photography and there is no concern for RCC [reinforced carbon carbon] or tile damage. We have seen this same phenomenon on several other flights and there is absolutely no concern for safety."

More than a week later, *Columbia* was set to return home. It was Saturday morning in Florida, and Steve Altemus had been coaching his son's soccer team, who were by then unlacing cleats on a grassy field not far from the Kennedy Space Center. As Steve headed for the parking lot, one of the other fathers shouted over, having just listened to a radio broadcast from the back of his truck.

"They just lost *Columbia,*" he yelled.

Steve stopped.

"What does that mean?" he replied. "What do you mean they lost *Columbia*?"

"They can't communicate with it. They don't know what's happening with it."

Immediately, Steve headed for Kennedy, just as another future Intuitive Machines engineer was in Houston monitoring the unfolding events. Rob Morehead, the company's future propulsion lead, had been working in the Johnson Space Center at United Space Alliance, a Boeing–Lockheed Martin joint venture designed to operate NASA's space shuttles. He had been monitoring consoles on the final leg of the trip. Everyone there was awaiting the shuttle's arrival. But as the craft passed over California and eventually Texas and Louisiana, communications started to fail. The heat sensors no longer responded. *Columbia*'s left wing seemed to be having problems. Two sensors meant to gauge hydraulic fluid temperatures were also not responding. More systems started to fail. The landing gear indicators were out. Then all signals stopped. By now, long-range radar should have picked up the craft's signals as the shuttle began its descent. But they just weren't there.

"FDO, when are you expecting tracking?"

"One minute ago, flight."

The last communication occurred at 8:59 a.m., when Colonel Rick Husband, the flight commander, was cut off as he responded to concerns about a loss of tire pressure data from *Columbia*'s left main landing gear.

"Roger."

Data transmitting continued for about five more seconds. Then nothing. Those in control appeared visibly tense, as evidenced by subsequent releases of the video. Minutes later, the mission operations rep, Phil Engelauf, received a call about contrails across the Texas sky. Amateur video would later reveal a single white streak breaking into multiple streaks.

An eerie quiet fell over the room.

"Lock the doors," said the flight director LeRoy Cain.

The seven astronauts on board had been killed. No one would be allowed to leave mission control until all data, reports, and firsthand accounts were recorded and stored. The accounting would need to begin immediately. A search-and-recovery effort was launched. More than 16,500 wildland fire personnel would be dispatched across thirty-eight counties in East Texas and fifty-two Louisiana parishes. Military planes would begin carrying NASA personnel and equipment to staging points all along the Arkansas border.

"I desperately wanted to be on the plane," Steve explained. Instead, he was tasked with leading the reconstruction effort at the Kennedy Space Center. Tens of thousands of pounds of wreckage were collected, loaded onto flatbed trucks, and delivered, where he would oversee the teams tasked with sorting and labeling. The debris had to be cataloged and recorded with barcodes, before being assembled onto a grid in a six-month effort to uncover the root causes of the catastrophe. The subsequent accident report found a culprit in that dislodged piece of foam, noting that it damaged the thermal protection system needed for reentry. The resulting hole in the tile allowed the superheated air to destroy the structural integrity of the wing, which eventually caused the entire craft to break apart. Investigators later blamed NASA's organizational culture for having "as much to do with this accident as foam did"—drawing comparisons to the earlier *Challenger* disaster, where O-ring seals had malfunctioned due to lower-than-expected launch temperatures. In both circumstances, engineers concerned about the safety of the vehicle had the burden of proving their craft unfit to fly, and yet were also hindered by the constraints of time, resources, and pressure to launch. Meanwhile, inside that fifty-thousand-square-foot hangar at NASA's Kennedy Space Center, teams of workers had assembled more than eighty-two thousand pieces of wreckage. Some of it discernible. Much of it a snarled mess, or "absolutely mangled," as the NASA administrator Sean O'Keefe described it. A caramel-like smell pervaded the entire facility, the charred remnants of insulation and melted metals that shocked the senses.

"We had gotten used to it," Steve explained. "But some people would walk in the hangar and break into tears. Other people would lose their breath, and had to run out.

"For us, putting it back together . . . was healing. It really had a personal effect of the deep sensitivity towards our loss and the loss of not only the crew, but all the thousands of workers that put that vehicle together."

NASA, meanwhile, would never be the same. The shuttle program would be suspended for two years as questions formed about the wisdom of the very program itself, with President George W. Bush later announcing that the program would be terminated. From its initial launch on April 12, 1981, to its final landing on July 21, 2011, it had carried out 135 missions. In those three decades, there had been two catastrophes.

That was enough to sap public support.

"We cannot find any justification to continue the deficit funding of a program that has no application other than proving that with enough money America can do anything," Bush explained. "The whole world knows that already, so why keep spending money on it."

A seismic shift, meanwhile, appeared to be stirring in the space market, with private companies like SpaceX and Blue Origin fast developing their own vehicles. In 2006, Boeing and Lockheed Martin joined forces to create United Launch Alliance, just as Virgin Galactic signed a twenty-year lease for a spaceport in New Mexico to continue developing its own suborbital craft. Just months after the *Columbia* tragedy, Elon Musk addressed an audience at Stanford University, calling the shuttle program "incredibly expensive and really quite dangerous." He then openly wondered "why we had not made more progress since Apollo." "That doesn't really jell with all of the other technology sectors out there," he added. "The computer that you could have bought in the early seventies would have filled this room and had less computing power than your cell phone. And so just about every sector of technology has improved. Why has this not improved?"

A better path, he surmised, would be to allow space activities to be "led by the spirit of free enterprise." Those words would later seem prescient. And yet state-run space programs around the world were also kicking into gear with an aura of ambition and urgency. A new race was now clearly unfolding, not just between nations, but over how the future itself might be defined.

· CHAPTER 17 ·

# MOON MOUNTAIN

A four-billion-year-old mountain looms over the Moon's south pole. More than 16,400 feet tall and basking in peaks of eternal sunlight, Malapert massif offers the prospect of both constant sunlight and an unobstructed line of sight to Earth for communications, just the kind of qualities a permanent human settlement might need. That also meant Moon landers would have to face the daunting challenge of touching down on—or at least near—a mountain that was taller than Mount Rainier. Yet if they could get there, and do so repeatedly, they'd score access to one of the Moon's most valuable pieces of sun-drenched real estate, close enough also to those shadowed craters where water ice was believed to be trapped.

It would also, however, be a spectacular place for an observatory, reflected Steve Durst, director of the Hawaii-based International Lunar Observatory Association. We had spoken briefly about it by phone, and it was clear that the idea had long occupied his imagination, a place where humanity could gaze more clearly into the universe from the edge of another world. Malapert wasn't fully on the lunar far side, but—with the right positioning—a future telescope could take advantage of the Moon's partial shielding from Earth's distorting radio noise, and thus offer more untrammeled views of the Milky Way. The problem was, however, Durst needed a space program focused on the Moon. And by the mid-2000s, there were no active American lunar missions in the works. China, meanwhile, was well into planning stages of its first phase of a lunar exploration program, with Beijing appearing intent on landing robotic explorers there within the decade. For someone like Durst, a trip to Beijing might have then seemed like an obvious decision. And why not? He had made inroads before on other international trips, particularly India's burgeoning space sector.

Two years earlier, during a conference in Bengaluru, Durst had in fact encountered a pioneer of the early Indian space program, U. R. Rao, who had proven instrumental in sending the nation's first satellites into space. Rao seemed to recognize the linkage between space technology and his nation's bid toward a more rapid modernization and technological self-reliance. Along with other Indian space visionaries like the physicist Vikram Sarabhai, the so-called father of the Indian space program, and A. P. J. Abdul Kalam, an aerospace scientist who became the nation's president, Steve Durst at times found himself mingling with top-flight Indian space leadership, who seemed to view space as both an arena for scientific discovery and a means of asserting sovereignty, particularly against regional military threats like Pakistan.

All of these connections could signal future collaborations. And if that could be done with India, perhaps a similar path might be possible in China, a country Durst had been visiting since the early 1980s. Over time, in fact, he had networked into the Chinese Society of Astronautics, closely linked to a main contractor for China's space program, the China Aerospace Science and Technology Corporation (CASC).

"I kept getting introduced to space technology people," he explained. "But I didn't have a concrete mission. In China, we found our mission. It's basically where things came together."

The more Durst talked about collaborating with upcoming Chinese lunar missions, the more excited he became. While in China, he was introduced to the same man who would years later call me in the middle of the night from Beijing. Xue Suijian, if you recall, had served as deputy director general of China's National Astronomical Observatories, and was the very man who had traveled to Argentina. Part of his mandate was building international collaborations.

To his credit, Xue still seemed to want to find ways to continue work within what would become an increasingly tense relationship between China and the United States, both for astronomy's sake and perhaps for his own deeper sense of the world beyond the borders of his own country. It was a perspective that may have been rooted in his own early surroundings, and that broader sense of history.

Born in the northwestern province of Shaanxi, bordered to the north by Inner Mongolia, he had effectively grown up at the beginnings of the old Silk Road, a network of ancient trade routes that spans fifteen hundred years and harkens back to the Han dynasty. These were relics of a very different time, considered among the most expansive and influential periods in Chinese history. In nearby Lintong District, an army of clay soldiers stand guard over the tomb of the Qin dynasty emperor Qin Shi Huang, who had reigned more than two hundred years before the birth of Christ. It was a history Xue had referenced with pride, not quite as a patriot, but more as a reminder of both Chinese potential and its legacy. "Most Chinese people are proud of their long history," he told me.

And yet in that moment, the tone of his voice shifted.

"What gives us a very different impression in China," he began, "was the humiliation in the latest two hundred years." It then seemed quite clear just where the conversation was headed. Again, even among its astronomers, there lingered references to the Opium Wars and a presiding sense of just how past vulnerabilities had been exploited by Western powers and Japan. For the younger generations especially—including its new crop of space entrepreneurs—that historical memory coupled with China's subsequent resurgence, particularly in space, seemed to shape an almost common sense of purpose and destiny.

· CHAPTER 18 ·

# THE MAKINGS OF CHINA'S COMMERCIAL ROCKETS

A modest apartment in a provincial capital might have seemed an unlikely home for the creator of what would become among China's most powerful commercial rockets. Changsha is a bustling metropolis of more than ten million residents, with so many streetlights and billboards, and such high humidity and pollution, that it all tended to obscure the night's sky. Yet not everyone deprived of the stars neglects to seek them. Astrophysicist Neil deGrasse Tyson once explained that he "had not seen the true night sky my entire life, having grown up in the Bronx," and that it was only during a visit to the Hayden Planetarium that his "lifelong quest to learn more about the universe" began. Perhaps in a similar way, Changsha—despite its urban glare—would nevertheless prove an effective incubator for the nation's space ambitions. Once governed by the rhythms of subsistence agriculture, the city had experienced a notable shift, with per capita disposable incomes rising from 7,400 yuan in 2012 to 18,300 yuan in 2021—at least according to official statistics. An e-commerce surge, supported by government subsidies and a growing emphasis on STEM education, had laid the foundations for a digital transformation.

But beneath that emerging veneer of economic progress and digital ambition were traces of an older and far more austere identity. The military history was unmistakable. And having spanned some three thousand years, it could not be easily discarded. Repeated battles had included four Japanese attempts to take the city during the Second Sino-Japanese War. And though it was Chinese Nationalists under Chiang Kai-shek who had largely defended the city, the role of the Communist Party was increasingly seen as central in China's fight against Japan. Changsha is, in fact, where Mao Zedong would attend Hunan Teachers College and become politically active. In 1927, Mao and other political leaders attempted to seize the city as part of the so-called Autumn Harvest Uprising. Unsuccessful, the effort nevertheless helped galvanize a peasant rebellion against a Nationalist government that ultimately led to its ouster. Years later, a mammoth, 105-foot-tall granite sculpture of a young Mao would be chiseled onto an island in the heart of the city. And it was that legacy that a young space entrepreneur named Yao Song would mention during our very first meeting.

Like with Xue, the meeting had been the product of my own prolonged attempt to unveil China's nascent quasi-commercial space sector. And given the rare nature of open discussions, especially with an American journalist, Song seemed remarkably candid, even when we first spoke. A young man in his twenties, he grew animated when the conversation veered toward his hometown roots and its military legacy.

"I think something in the culture of our Hunan province that you find in Chairman Mao and a lot of leaders in early China is that over 50 percent of the generals when they established the [People's Republic of] China were from Hunan province," Song explained. While it's not clear if those figures are accurate, Mao's chief of staff Su Yu and General Peng Dehuai, who served as the country's early defense minister under Mao, both hailed from Hunan—figures whose stories were a part of the cultural and historical touchstones that had shaped Song's formative years. In fact, in one of our several conversations over the course of nearly two years, Song would describe himself as a "crazy military fan," especially during those early years. And as China's rapid military mobilization got under way, he had become enamored with "ships, missiles, and tanks."

"I'm truly addicted to all these things," he told me, apparently a natural fit, given the nexus between the PLA and space. And yet for all his fascination, he also seemed drawn to deeper, almost existential questions, which tended toward humanity's coming place in the cosmos, and the role this new generation of scientists and engineers might play in shaping it.

"My dad bought a lot of books about the universe," Song explained to me one evening in the spring of 2024. At age four, "he'd read to me [Stephen Hawking's] *A Brief History of Time,*" the kind of early exposure that may have sparked a broader curiosity about the nature of the universe and the laws that governed it. As the years progressed, Song would study electronic engineering, focusing on chip development and superconductors, while the rocket company he dreamed of creating would, for now, have to wait. Early on, it was neural networks that captured his attention, deploying artificial neurons across multiple layers to extract hidden patterns in data. These complex systems reflected a kind of scaffolding for machine learning, enabling them to adapt, and even approximate certain aspects of what arguably resembled a degree of artificial cognition. At just twenty-two years old, he began working on "face recognition and other AI application" needs developing across an early Chinese start-up landscape, utilizing high-performance chips and computational power that hinge in large part on hardware, the optimizing of algorithms, and the leveraging of distributed computing systems.

From Song's vantage, he saw room for improvement—particularly in a technique called deep compression, in which the overall size of the network and its underlying hardware could be effectively reduced without compromising performance. In this case, the focus was on a common aspect of that technique called quantization, which essentially constrains how input values are mapped with a set number of elements, thus reducing the inference time of how long the AI needs to make decisions.

"The bottleneck for computation for neural networks is the bandwidth," Song explained. But with this process, "we can reduce [dependency on bandwidth for] the memory access by maybe 95 percent, or even more."

Part of the idea was that ever smaller chips could extend Moore's law, even though innovations in quantum computing and neuromorphic computing can create scenarios for increasing computational power that do not rely principally on scaling. Still, nanofabrication seemed to offer profound new opportunities, engendering breakthroughs and supercharging China's tech ascendency in everything from data analysis and autonomous navigation to robotics and military planning.

Song, however, needed to put his ideas to use.

So, in 2016, after meeting with his old professors, he decided to co-found an AI company called DeePhi Tech, focusing on deep learning acceleration hardware and software, which could be applied in drones, data centers, and surveillance. DeePhi Tech, it turns out, is a combination of the words "deep" and "philosophy," a nod at Song's sense of exploring the almost philosophical principles behind AI while also driving its advancement. Within two years, however, in July 2018, the company was acquired by San Jose–based semiconductor firm Xilinx for about $230 million.

The timing here is important, given U.S. efforts to ban Chinese IT systems, namely Huawei, were due to take effect a month later, in August of that same year. Concerns over Chinese espionage were mounting. But by then, Song had already secured the funding he needed to pursue a lifelong dream: A space company of his own.

He was about to start building rockets.

· CHAPTER 19 ·

# IMAGINING TOMORROW

The year was 1966, the start of the final full decade of Mao's rule. After a ruinous economic campaign bent on rapid industrialization, which had contributed to one of the worst famines in modern history, the unity of his Communist Party now seemed in question. Between 1959 and 1961, as part of Mao's so-called Great Leap Forward, some thirty million people would starve to death as political tensions and ideological debates simmered and the beginnings of new and potentially powerful political factions took hold. To shore up control, Mao launched his Cultural Revolution, in large part to destroy his enemies and purify party ranks. What happened next is considered one of the most violent periods in modern Chinese history: a mass mobilization that resulted in millions persecuted and imprisoned under the guise of strengthening the nation's Communist fervor.

That was the backdrop for the first of an esteemed Chinese sci-fi trilogy, and one of the most formative novels in Song's early life. Published in 2008, *The Three-Body Problem* by science fiction author Liu Cixin involves a young scientist witnessing the brutal public persecution of her father, a physics professor, who is beaten to death by Red Guards. Traumatized and disillusioned, she nonetheless works for the Chinese military on a secret government effort to bounce signals off the Sun as an amplifier and into the depths of space. Eventually, the signal draws a reply from an advanced civilization seeking refuge from their home—a triple-Sun system. Known as a three-body problem in celestial mechanics, where competing gravitational forces work to create unpredictable and chaotic orbits, the system inevitably wreaks havoc on their civilization, breeding instability and the desperate pursuit of a new home to overtake. Rooted in scientific principles, the trilogy gained immense popularity in China—with the Chinese streaming service Tencent streaming an adaptation—even as the story offered critical portrayals of the Cultural Revolution. Remarkably, a story homegrown in China had sparked controversial questions about the role of technology in sustaining authoritarian rule, while also introducing frontier physics concepts, such as quantum entanglements, which could theoretically allow for instantaneous interstellar communications. The influence of sci-fi in igniting curiosities and scientific possibilities was yet again at work, with certain undercurrents of political commentary. And for Song, when he read it, the effects were profound.

"I think *Three-Body Problem* inspired a lot of young kids in China who became devoted to go to outer space," he told me from Beijing.

"I was born in 1992 . . . [Back then] we didn't regard aerospace as a major working area. We didn't think there were jobs and opportunities in this area. But once *Three-Body Problem* came out, there were so many fans. They were just jumping into aerospace, jumping into AI area, or other high-tech areas.

"I think it really was an inspiration for a lot of people."

Within a few years of graduating, Song (by now age twenty-four) would be named to MIT's Innovators Under 35 China list, an annual grouping of leading pioneers and innovators across the country. His ambitions were also clearly growing. DeePhi was ultimately the reason for the MIT honor. But the name, Deep Philosophy, also betrayed what he considered a core, almost metaphysical aspect to his work, related to a broader scope of learning, intelligence, and human-machine potential, with much of it centered on the unlocking and exploiting of patterns.

"We are all connected. And there's a very huge complex graph, connected by dots," Song explained. "The dots are single beings to single beings. So, there's a code to society." The idea in some ways envisioned Earth evolving into a kind of planetary supercomputer; a vast network of solar-powered sensors, AI, and autonomous systems linking continents and orbits, and continuously processing data on unprecedented scales.

Perhaps in the way that ants, whose individual cognitive capacities are relatively small, nonetheless collectively employ a kind of "swarm" intelligence through the sum of their collective actions in order to engineer mounds, build intricate tunnel structures, promote hierarchical relationships, and even formulate highly organized systems that locate food sources and avoid predators, so too might humanity be capable of engineering something far more complex by aggregating the sum of its parts. It is a concept that perhaps at first blush might seem more aligned with Eastern philosophers like Confucius who tended to emphasize social harmony, hierarchy, and the collective good, as opposed to classical Western Enlightenment thinkers who focused more on individual liberties, freedom, and autonomy. Through that prism, Song—now a leading Chinese space entrepreneur—seemed to be alluding to broader cultural and historical differences as they relate to humanity's coming place in the cosmos. This race in space, and the technology behind it, suddenly seemed to be also driving a philosophical competition in the guiding principles for how a future humanity might structure and govern itself: the perceived good of the collective versus the rights of the individual. These, of course, were broad generalizations. In fact, during a follow-up conversation, Song even seemed to qualify those very ideas.

"I think we cannot be a swarm group. We are human beings. We are singles. We are individuals," he explained. "But I think [technologies] can help better express ourselves and help everybody to understand each other much better." Still, if new mega satellite constellations could effectively be leveraged to communicate directly with growing arrays of space-based and stratosphere-based platforms, new interconnected and blindingly fast tech ecosystems seemed also likely to emerge. How systems might then be integrated into broader governance principles and rationales seemed like incredibly important questions.

Yet without a crystal ball, they were also profoundly difficult to answer. Perhaps that is also why conversations about whether humanity *should* pursue certain advancements are so often overshadowed by the momentum of science and engineering, whose focus tends to be on merely *how* those advancements might be achieved, rather than *whether* they ought to be.

Song, to be sure, reflected on such ideas. But ultimately, he was a builder.

And to him, some of the more profound potentialities of these new AI-fueled developments lay in the Internet of Things, where everyday physical objects communicate and exchange data to afford digital monitoring and near-unprecedented control over the physical world. Picture, for instance, satellites autonomously sending signals to each other to optimize coverage; rovers and landers communicating in real time about atmospheric pressure, soil composition, and radiation; and mining operations capable of monitoring the extraction and transport of mineral resources on asteroids or beneath the lunar or Martian surfaces. The implications for space infrastructure were immense. Off-Earth habitats could improve logistics and inform robotic assistants to carry out otherwise dangerous repairs, maintenance, and reconnaissance in areas unsafe for humans. The limitations of our own biological vulnerabilities, in that respect, were becoming painfully clear. The age of the human explorer had its limits. Far more expansive possibilities now rested with our machines, particularly in the realms of the distant solar system and interstellar space.

If self-replicating probes could, for instance, be sent out as scouts to the distant corners of the Milky Way, and at fractions of light speed, what might we find? And how might those machines themselves develop and replicate so far from human oversight? What unforeseen consequences might emerge? The possibilities seemed both endless and unknowable. Of course, the divide between human and machine explorers didn't have to be a strictly binary one. New possibilities were emerging with human augmentation and brain-computer interfaces (BCIs). Basic concepts like wearable tech for health monitoring and bioenhancements were just a starting point, and more advanced functions like the development of exoskeletons that could help astronauts perform physically demanding tasks, the development of virtual and augmented reality training scenarios, or teleoperation, which provides real-time feedback and visual displays of a robot's surroundings as it operates outside the habitat, were all fast in development. Biotech implants might then be a next natural progression, with the human body serving as the principal interface, and thus accelerating and reshaping how humans digest and exchange information. Advanced data processing that served as cognitive support, while bridging gaps like language and cultural barriers, might then be just the beginning.

It could even help with dating, Song explained with a chuckle.

"A lot of engineers don't know how to talk with other people," he said, "especially cute girls."

As we talked, notions of what might have once seemed pure science fiction had somehow taken root among this new breed of algorithmically armed thinkers and tinkerers. Could brain-computer interfaces, for instance, one day allow for direct thought-to-thought communications, bypassing traditional forms of communication like speaking or typing? Could the volume of information exchanged then grow exponentially between parties? Could such systems be augmented with virtual reality, allowing for immersive experiences, especially when it came to interplanetary and deep space exploration, where extreme conditions were hazardous to human biology? Taken a step further, would it then be possible for BCIs to effectively revolutionize collaborative, multidisciplinary research, especially when applied in tandem with AI? And what sorts of privacy and nefarious actor concerns might emerge when brains were directly interfaced with technology? How might we need to reconsider freedom of thought? Could a brain be hacked? How might brains become interoperable with other physical devices and objects? And would a hierarchical rift begin to emerge between those with and those without such implants?

Certainly, of course, not everyone was convinced.

Those like Andrew Jackson, professor of neural interfaces at Newcastle University, who in 2005 helped develop autonomous implantable computer chips for neural recording and stimulation in primates, had expressed doubts about what those like Song and, of course, Elon Musk had endeavored.

"Musk has more ambitious goals, hoping to read and write thoughts and memories, enable telepathic communication and ultimately merge human and artificial intelligence (AI). This is certainly not 'trivial,' and I don't think the barriers can be overcome by technology alone," he wrote in September 2020. Referencing Neuralink, a neurotechnology company founded by Musk and a team of eight scientists, Jackson continued,

> Today, most brain-machine interfaces use an approach called "biomimetic" decoding. First, brain activity is recorded while the user imagines various actions such as moving their arm left or right. Once we know which brain cells prefer different directions, we can "decode" subsequent movements by tallying their action potentials like votes.
>
> This approach works adequately for simple movements, but can it ever generalise to more complex mental processes? Even if Neuralink could sample enough of the 100 billion cells in my brain, how many different thoughts would I first have to think to calibrate a useful mind-reading device, and how long would that take? Does my brain activity even sound the same each time I think the same thought?

Four years later, Neuralink had developed implantable brain-computer interfaces. And thus, what was perhaps most interesting was that these ideas were now shifting from theory to more practical considerations. The technology had, in fact, seen marked improvements. In January 2024, Neuralink had made history by successfully implanting a chip in a thirty-year-old quadriplegic Arizona resident named Noland Arbaugh, who had been paralyzed after diving into a lake. Using pattern-recognition algorithms to decipher neural activity correlated with movement and sensory perception, mental commands apparently could then be used to control external devices, such as a robotic limb or even a laptop. If Arbaugh merely thought about moving the cursor on his computer, his action was seemingly enough to make it happen, raising all sorts of speculation about the technologies' uses in space.

It was a remarkable pace of advancement, with even the fundamental parameters of how machines were defined seemingly changing. Computers themselves soon might not even look and feel like computers. A Swiss company called FinalSpark, for instance, sought to move beyond silicon-based hardware and employ synthetic biology to construct computer architectures powered by brain cells in a system called Neuroplatform, which enabled "researchers to run experiments on neural organoids with a lifetime of even more than 100 days," according to a paper published in the open-access, peer-reviewed journal *Frontiers in Artificial Intelligence*. The goal was to devise architectures that harness the processing potential of living brain cells, rather than conventional microchips, by employing three-dimensional, lab-grown clusters of brain cells that mimicked certain functions of the human brain. In the broader scope, that suggested a future in which hybrid biological-digital systems could complement or even surpass traditional methods, at least in certain applications such as AI training, neuroscience research, or ultra-low-energy computation. In the context of space, the implications of long-duration space travel were hard to ignore, whereby biologically integrated systems might afford more intuitive human-machine collaboration and autonomous problem solving, while self-healing systems could adapt to problems like the degrading effects of radiation. Musk, for his part, had described such concepts as endeavoring to "achieve a symbiosis with artificial intelligence."

It was a striking statement.

And Song, like much of the Chinese tech and aerospace industry, had been paying close attention.

To him, Musk was the model, albeit with some important caveats.

"I truly appreciate what [Musk] did," he said. "But I'm not that crazy as him. For example, he will say, I'm going to Mars in early stage of SpaceX. In China, if you are going to say this, you get laughed at by a lot of people, a lot of people." Still, Song's early investments after the sale of DeePhi Tech nonetheless seemed to mimic those of the South Africa–born billionaire.

"I did some research for gene editing, for brain science, for neural interface, really for drug discovery with some new AI algorithms," he explained, investing in a company "similar to Neuralink," a Chinese biotech start-up called NeuraMatrix that endeavored to develop neural interfaces implantable in the human brain.

"What is consciousness? What's that? And how is memory stored in our brain? I think this is a really interesting area to discover," Song later told me. But, like Musk, he was also focused on those more tried-and-true mechanics of getting to space.

He wanted rockets.

And so, in a bid to compete with his model, Song would co-found a launch company of his own.

"There is a huge opportunity for commercial rocket[s] to send many satellites to outer space," he explained of his decision to form Orienspace in 2020, employing $150 million in funding that he and his partners were able to secure in part from the Chinese venture capital firms Sequoia China and Matrix Partners China.

For him, China's launch sector needed more of the entrepreneurial infusion he had seen in the United States, where he had studied only briefly. Many of the "previous commercial rocket companies [in China were] run by previous government officers. I didn't think they had a really good sense of . . . how to reduce the cost [and] how to improve efficiency."

For a young, rich, and already successful Chinese entrepreneur, enamored with military tech and his country's thousands-year history—and who also happened to be coming of age at a time when his nation was doubling down on space, quantum technologies, and AI—the timing could have scarcely seemed better. His company would emerge as one of China's champions when it came to Beijing's broader ambitions in the cosmos, even if SpaceX was still light-years ahead. That gap in capability, however, didn't stop those in China from watching closely. Across the Pacific, in fact, one of those SpaceX rockets was now preparing to make history.

The race to the Moon was now entering a new phase.

· CHAPTER 20 ·

# THE NIGHT BEFORE BLASTOFF

On a cool evening in February 2024, a sliver of moonlight bathed the Kennedy Space Center in a soft glow. In two days, more of the Moon would appear with more of a radiance, offering greater visibility as to the location of Intuitive Machines' new landing zone. The site, adjacent to the 190-mile Malapert A crater, was a relatively flat region compared with much of the pockmarked lunar south. Named after the seventeenth-century Jesuit astronomer Charles Malapert, who used his observations of comets and the stars to falsely refute the heliocentric hypotheses of Galileo and Copernicus, the location was low on the lunar horizon and relatively close to the site of the planned Artemis Base Camp.

Now the only thing left to do was get there.

Steve was in the lobby of the Hilton Garden Inn in Cocoa Beach. Tired, and flanked by two engineers, seemingly out of breath, he was headed for the pad. This was a ritual, a routine he had developed back in those early shuttle days, when—in the still of the early evening—he would talk to the vehicle, coaxing it as if it were alive.

"You behave," he would whisper to *Odie* that evening.

"Rockets," he later explained to me, "have their own language."

*Odie* was the company's lunar lander, roughly the size of a British phone booth, that would soon ride atop that SpaceX rocket and into orbit. It had gone through countless tests. But many here knew that the first big test had centered on that new methalox engine and whether it would light.

SpaceX launch facility at the Kennedy Space Center, 2024.

Ignition had been tested in a vacuum chamber. But to really know, it had to just do it. More than a dozen engineers had been dispatched to Hangar AO at Cape Canaveral Space Force Station, now used predominantly by SpaceX. Away from their families, and working out of what was effectively a bunker, the team was charged

In the nights before that historic Intuitive Machines IM-1 launch on February 15, 2024, executive leadership put on a series of events. Recording my observations of the team as launch time neared.

with "knocking out gremlins in the system," Steve explained, subsisting as they had been on a steady diet of Krystal hamburgers and donuts.

"They were looking pretty ragged," Steve explained.

Meanwhile, the hotel was alive with activity. Investors, contractors, and the well-heeled mingled with members of the Intuitive Machines and SpaceX teams with less than twenty-four hours to go before liftoff.

These were the last checks.

"Will the engine work? Will it fire?" Steve wondered. "There's no more testing we can do."

Meanwhile, NASA officials in Washington and those at the White House quietly wondered. After that *Peregrine* mission, the agency could ill-afford another failure. Further delays could also jeopardize America's lunar ambitions, just as China's Chang'e 7 mission—a robotic lunar lander—geared up for what was then its scheduled 2026 launch. Days earlier, Steve had spoken with NASA's associate administrator Jim Free, who had led the agency's Explora-

Steve Altemus, co-founder, Intuitive Machines, Kennedy Space Center, before the launch of IM-1, in early February 2024.

tion Systems Development Mission Directorate and oversaw both the Artemis program and the planning of NASA's Moon-to-Mars architecture.

"I want to look you in the eye," Jim said. "Are you ready to go?"

"Yes," Steve replied.

By then, of course, what else could he really say? What choice did he have? And as such, Steve wouldn't get much sleep that night, waking up around 3:00 a.m., much to the chagrin of his wife, Brunella, who had flown with him to the Cape from Houston. In fact, much of his family had joined them in the festivities at Cocoa Beach the night before. Some washed down their steak and chicken with margarita-filled, Intuitive Machines–branded mugs. Others were going over last checks.

"The team is exhausted," Steve said, addressing the crowd. But, he added, "they're going to do the job that the United States is looking for them to do," pausing for effect before pointing skyward.

"I want everyone to look to the left. You'll see a beautiful and sacred perspective of the Moon. It's a waxing crescent Moon. And you can see the south pole of the Moon lit up . . . and very soon in the future, it will give back to us a little glint of life of the Intuitive Machines Nova-C lander, reflecting on the surface for over the next billion years."

Launch was now just hours away.

Meanwhile, the company's propulsion lead, Rob Morehead, was staying with his family at the Hilton Oceanfront, a few miles down the road. I hadn't seen him since those early days on the Houston tarmac, when he was still test-firing the engine. By now, like everyone else, he also seemed a bit out of breath and, naturally, second-guessing.

"Honestly, in hindsight, we shouldn't have put so many firsts onto the vehicle," Rob told me. "But these were things that were on the edge of being good enough when I left NASA. And so it felt like it was reasonable to push them all forward."

To reduce mass, the craft employed linerless tanks and—again—elected to use a continuous thrust propulsion. This is just what Rob had been trying to figure out in Houston, poring over data to get just the right injector configuration and mix of liquid methane and oxygen. But, of course, in microgravity, the behavior of propellant floating inside a tank could be unpredictable, causing fuel to flow in the wrong direction.

"When you turn the engine back on . . . it could blow up," he said.

"That's one of the suspected reasons why some of the previous landers have crashed," Rob went on. "We have a deep throttling engine that we never shut off . . . burning the engine at low-power levels."

But there were other concerns as well.

Methane as a rocket fuel requires cryogenic storage to keep it in liquid state. As the tanks are fueled, part of that fuel is also simultaneously boiling off. The process, therefore, has to be done quickly to prevent excessive fuel loss, while the tanks themselves need to be specially insulated to keep the methane as a liquid throughout the journey. The team had been busy inspecting SpaceX's modified pad, along with their own system's valves to inspect for leakages.

It was a delicate operation.

But just as those physical systems were sorted, an entirely new kind of threat seemed to be stirring. It wasn't mechanical. It wasn't even visible. A cascade of unexplained network problems had suddenly emerged.

Rob explained it to me cautiously.

The team couldn't be sure where they were coming from. But the effect was now hampering connectivity between ground teams and the vehicle itself, he explained. There were "a bunch of network issues." Without a reliable exchange of data between the vehicle and ground operations, the team might struggle to monitor *Odie*'s systems and make real-time adjustments. In the moments that followed, speculation simmered. Was this a technical hiccup or something more deliberate?

"We have a whole lot of software constants that tell us what to do at certain times," he added. The craft relied on a private lunar telemetry tracking network, replete with five data relay satellites and a series of ground stations around the world that provided line-of-sight communications. But data security was still a vulnerability. As the team worked the issue, questions about the software's integrity began to surface. "We were working a lot of network issues, [and] some outside actors [were] apparently messing with the network," Rob noted. And that, he added, seemed to be "messing with the data." It wasn't clear who exactly those "outside actors" were, and his comments could not be independently verified. But in a way, this sort of thing might have been expected.

When it came to the space race, fears of a cyberattack were ever present.

In fact, in November 2023, Japan's space agency reported being hit with one such attack in which there "was a possibility of unauthorized access by exploiting the vulnerability of network equipment." In March of that same year, a notorious criminal ransomware gang called LockBit taunted Elon Musk by threatening to release details of his rockets, claiming to have stolen thousands of schematics from one of the company's contractors. In a message posted on the dark web, the gang apparently addressed the SpaceX CEO directly. "Elon Musk, we will help you sell your drawings to other manufacturers, build the ship faster and fly away," the post said. "We will launch the auction in a week." By August, the FBI, the National Counterintelligence and Security Center, and the Air Force Office of Special Investigations had even published a two-page advisory that focused specifically on space.

The threats, it seemed, were growing.

"Foreign intelligence entities (FIEs) recognize the importance of the commercial space industry to the US economy and national security, including the growing dependence of critical infrastructure on space-based assets," the advisory said. "They see US space-related innovation and assets as potential threats as well as valuable opportunities to acquire vital technologies and expertise. FIEs use cyberattacks, strategic investment (including joint ventures and acquisitions), the targeting of key supply chain nodes, and other techniques to gain access to the US space industry." The report added that such spy efforts were intended to siphon intellectual property to the benefit of their own space agencies, allowing them to effectively leapfrog innovations it took U.S. firms time and resources to produce, thereby reducing their competitiveness on the global market. The agencies further warned of foreign entities collecting sensitive data related to satellite payloads, with the intention of both degrading U.S. abilities to provide critical services—particularly during emergencies—and identifying vulnerabilities in emerging commercial space infrastructure.

Reaching the Moon was hard enough. But now, the prospect of disruptive, outside hands at work seemed to be further complicating the mission. A few hours later, a few miles from the pad, Ray Kowalik had been sipping his drink. The former Burns & McDonnell CEO had been among those attending the prelaunch event, having played a major role in building and designing Intuitive Machines' nearly 106,000-square-foot operational hub, equipped with mission control rooms and propulsion test sites.

Security, however, was top of mind.

"I guarantee you someone in here is trying to get in," he said to the group that had gathered at that prelaunch event. SpaceX and NASA had run multiple background screenings about who precisely was allowed on-site, not that anyone would need to be for an effective cyber strike. Meanwhile, countdown was approaching. And tensions were high, not just for those in that room, but for the global audience that had tuned in to watch the company's video streaming of the night's events, including both Xue and Song in Beijing, the latter of whom was fresh off the inaugural flight of his own company's Gravity-1 rocket.

In China, it was a moment that had carried a different kind of weight.

· CHAPTER 21 ·

# LAUNCH FROM THE EAST

The skies above the Yellow Sea were still dark when the first crowds began gathering along its western shores. They were there to witness history. It was the morning of January 11, 2024. And out in the water, atop a large barge, was soon to be a demonstration of the most powerful commercial launch vehicle a Chinese company had ever assembled—a high-stakes moment both for Song's company, Orienspace, and for Chinese commercial spaceflight.

Standing at roughly thirty meters tall with seven solid rocket motors, the Gravity-1 boasted a six-and-a-half-ton payload capacity to low Earth orbit. Although it was considerably smaller and less powerful than those rival Falcon workhorses, the goal was ultimately to close the gap with SpaceX. At the moment, it was bobbing up and down near the coastal city of Haiyang, where China's main space contractor had been developing the country's latest spaceport. With the ability to launch up to thirty satellites at once, Gravity-1 now seemed like a strategic asset, one that could better carve out Beijing's own place in orbit. That kind of capacity now seemed especially critical as China accelerated deployment of its AI-enabled satellites and the emerging Three-Body Constellation, designed to process vast streams of data directly in space.

Meanwhile, China's quasi-commercial space industrial base was expanding. In Yizhuang alone, a planned innovation hub southeast of Beijing, more than twenty commercial launch companies were designing their own launch vehicles, including Galactic Energy and iSpace, along with numerous satellite manufacturers, as well as companies devoted to AI, semiconductors, and other forms of advanced manufacturing. The idea was to concentrate adjacent technologies within tightly integrated clusters, where cross-sector collaborations and state-backed funding might help China leapfrog its rivals. SpaceX's reusability and the sheer scale of its Starlink constellation was still the benchmark. But China's emerging constellations, including Three Body and its GW Project, or Guowang (or national network), now represented a calculated effort to counter with its own megaconstellations. The broader goal, according to the associate professor Xu Can with the People's Liberation Army's Space Engineering University in Beijing, was to "prevent the Starlink constellation from excessively pre-empting low-orbit resources."

The problem, however, was scale. To really compete in orbit, these rockets would have to be reusable. "For China to establish its own Starlink, the key is to master reusable rocket technologies," emphasized Qu Wei of the China Academy of Aerospace Aerodynamics in Beijing.

And yet, for the moment, the country's space sector wasn't quite there. Chinese companies like LandSpace, Space Pioneer, and Orienspace all had reusable rockets in development, along with state-run Long March 12A, but none were ready for prime time. Still, ambitions were rising, not just toward feasibility, but also in the economics of launch, which directly informed Chinese capacity to put more compute in orbit.

Song's hope was to eventually fly nearly a hundred reusable rockets annually to support those constellations, undercutting SpaceX's launch and payload costs, even as the nation's broader satellite deployments remained woefully behind schedule; largely for want of a reliable, reusable launch vehicle. Musk's then Hawthorne-based company by then had roughly seven thousand Starlink satellites, and aimed to eventually deploy forty-two thousand. More concerning for China, perhaps, was Starlink's national security derivative, Starshield, which aimed to provide secure communications, as well as target tracking, reconnaissance, and surveillance, to U.S. national security sectors—seemingly in line with the Trump administration's later push of its "Golden Dome" missile defense shield.

"Our target is to go lower than Falcon 9," Song told me. For context, a SpaceX Falcon 9's payload cost at the time just over $2,700 per kilogram, compared with $20,000 per kilogram by expendable means. A comparable Galaxy-1 price tag was roughly $4,000 per kilogram, largely because it wasn't reusable. Plans for the Galaxy-2, however, were meant to change that, as were at least six other Chinese companies, including Space Pioneer, which by 2024 had added $207 million in a blend of private equity and state-tied investment for a reusable medium-lift launch vehicle known as the Tianlong-3. China's main state-owned aerospace company—China Aerospace Science and Technology Corporation—not to be outdone, had unveiled plans for its own superheavy-lift launch vehicles, including the Long March 9, which had a strikingly curious resemblance to the SpaceX Starship.

The difference lay in the approach.

Where SpaceX focused on rapid innovation and iterative design driven initially by private capital, China's state-led, large-scale infrastructure and manufacturing support focused on building its industrial base. And yet, in truth, each enjoyed national support. Whereas Chinese authorities tended to also support the supply side of a company's development, which included the building of factories, companies like SpaceX enjoyed massive federal contracts as service providers. Chinese companies, therefore, may be harder to start, but once established, those firms stood to benefit from immense and patient sources of capital, backstopped and overseen by centralized authorities. This quasi-commercial shift seemed to begin in earnest in the mid-2010s. Before then, China had almost exclusively relied on state-owned entities. And yet leadership now seemed to recognize the importance of commercial ecosystems that could initiate innovations at a more competitive tempo. Authorities were looking to those like Song to fill the gap, with a clear push toward dual civilian and military applications. To succeed, particularly on the military ledger, it had to be able to launch within twenty-four hours of an "emergency" mission request. And yet, for now at least, Song's rocket had yet to reach orbit.

Standing three kilometers away from the launchpad, meanwhile, he was starting to get nervous. Until that moment, Song had been relatively confident about his company's inaugural launch, and had even invited colleagues and former professors to attend. But as the countdown began, his stomach churned.

"Once you hear 'ten . . . nine . . . eight . . . seven . . .' you get really nervous in your mind." And yet as he described the events, and where he was positioned, something very clearly didn't add up.

Finally, I asked the question that now seemed obvious.

"Were you in the launch control center?"

"No," he replied. "I was at the stage. There's a stage to watch the launch with thousands of people. The control center in China actually requires high-level confidential qualification. And I was not enough to get into that control center."

"Really?" I replied. "So this is your own rocket and you couldn't get in?"

"Yeah," he responded. "A lot of the operation actually is conducted by the military, not by us. This area is not a mature industry in China. So, once you get to launch your own rocket, there actually will be some technical experts to check everything. And there will be some military officers to check every process, doing tests, and finally counting down to see everything, and see whether it's a red light or a green light." It was a remarkable revelation. For all of China's commercial growth, there was still a clear limit to private enterprise and the limited sense of autonomy it could glean from the state.

Song, the company's founder, would have to watch from afar as a spectator. Minor delays ran through the morning, as he watched in anticipation.

Then it happened.

As the countdown ticked to zero, two massive plumes of gray and black smoke could be seen bursting from either side of the vehicle. The goal was to deliver into orbit three Yunyao-1 commercial weather satellites. But "for the first three or four seconds, you couldn't find the rocket," Song explained, referring to the exhaust and water vapor that by now had fully enveloped the vehicle. "I was really in a panic. Those seconds felt like a thousand years. But then I saw the rocket go straight up through the clouds and into space.

People were applauding, and they hugged me. And then I heard the voice from the broadcast say clearly, stage one . . . stage two. Eventually, I heard that the launch mission was completely successful. It was a huge moment." The rocket had also hit a financial milestone by becoming the "lowest launch cost in China, especially for commercial," Song explained, quickly pivoting to again discuss the company's broader purpose, and the need—as he put it—to "expand the circle."

This was a recurring concept for him, and relates back to that broader sense of the existential that always seemed to be lingering in the recesses of his mind. The so-called circle, in his estimation, was a reflection of Earth's finite resources and the limited availability of habitable territory relative to population growth and increasing energy demands. A wider circle, derived through humanity's spaceborne expansion, he posited, would therefore ultimately translate into more land and resources to effectively support that growth. For him, that was a compelling enough case for the kinds of launch capabilities and infrastructure that might ultimately yield orbital data centers, asteroid mining, and settlements atop other celestial bodies, which included the Moon and the race to it. Humanity would depend on it. If, however, the circumference of that circle remained fixed, and competition for terrain and resources continued to swell, "there will definitely be a war," he predicted.

· CHAPTER 22 ·

# COUNTDOWN

Back in the U.S., it was around 11:00 p.m. when a two-bus convoy wound its way through tall grasses and distant towers in the Florida wetlands. Unlike the fanfare surrounding Gravity-1's morning launch, night launches are more curious things. Against the dark, the rocket's rumble and intense glow can make it almost feel like daylight again, until the burn fades and the rocket becomes just another pinprick across the night's sky. Aboard that bus, it was quiet. Passengers fiddled with their phones, or just stared out windows. The tension seemed almost palpable, sharp and inescapable. After years of preparations, *Odie* was now ready for the Moon. As the two buses slowed and eased into a parking lot at the Kennedy Space Center, we funneled into Operations Support Building II, where—from the upper floors—*Odie* could be seen, illuminated in the distance. Inside, a table of cheese plates and chocolate hors d'oeuvres, alongside an open bar, did what they could to ease the nerves. Steve Altemus, meanwhile, was making the rounds.

It was an hour until liftoff when he approached a lectern, situated inside.

Everyone took their seats.

Steve then smiled a half smile.

"The last time I stood at this podium was a long time ago during a shuttle flight readiness review. The only reason you get to stand at this podium is because your system isn't behaving properly."

The crowd chuckled.

"Fortunately, that's not the case tonight. All systems are go.

Intuitive Machines co-founder Steve Altemus at Operations Support Building II at the Kennedy Space Center prior to launch of IM-1.

Moments after the successful Falcon 9 launch carrying IM-1 into orbit and on its way to the Moon in a bid to make history as the first commercial craft to ever touch down on the lunar surface. Intuitive Machines co-founders Kam Ghaffarian (*far left*) and Steve (*second from right*) and Brunella Altemus (*fourth from right*).

"For all of you that are here tonight," he added, "I really appreciate, and the whole company appreciates, your energy, your positive tone, your love, and your respect for Intuitive Machines and what we're doing. . . . I can feel the energy of so many people putting positive thoughts out there."

When he finished speaking, he and Kam Ghaffarian stepped out onto the balcony to exchange a final few words before blastoff. Remaining propellant had just been loaded into the vehicle. And as the final countdown began, the crowd started counting.

"Three . . . two . . . one . . ."

Nine Merlin engines ignited, sending out plumes of white smoke as a red-hot glow lit the night's sky. For a moment, the rocket seemed almost suspended in air. Then it gradually picked up speed and shot skyward, a low rumble and vibration rippling toward the balcony where all of us stood. Kam put his arm around Steve as the two founders watched *Odie* disappear, twisting its way into Earth's upper atmosphere.

The IM-1 mission had begun.

Stage separation would soon allow the second-stage booster to carry the lander into orbit. Tim Crain, meanwhile, was in mission control, staring, arms folded, at screens that were now peppered with a flurry of changing numbers.

"I want to make sure comms is happy with it," an unidentified voice rang out from mission control. "Comms?"

"Umm . . . standing by," responded the team's radio frequency communications engineer Micaela Landivar. Her voice seemed shaky. "I don't see packets updating in packet counts, unless I'm looking in the wrong spot."

Another moment passed before anyone in control said anything. A lack of "packet count" could suggest a communications problem. Everything from the spacecraft's power levels and temperature to the mission-specific progress had been baked into data that presumably by now should be transmitting.

"I'm not seeing anything populating in AOS [acquisition of signal]," she added. Another few seconds passed. A collection of held breaths seemed to punctuate the silence.

"All right, I'm seeing packets," Landivar said, a smile creeping up one side of her face. "It looks like we are seeing most everything we would expect as far as downlink is concerned."

That relief, however, would be short-lived.

No sooner had it separated from the second-stage booster than the first signs of trouble appeared.

*Odie* started to tumble.

That part was expected. On a timer, a host of functions were now supposed to spring to life, including the craft's flight software, radios, and thermal control. To stabilize the vehicle, however, *Odie*'s guidance navigation and control would have to power on. Helium thrusters could stop it from rolling and spinning. But to do so, the craft had to employ a system called star tracker, which worked by surveying the surrounding star-scape to pinpoint precisely where and how the vehicle was oriented. With that information, *Odie* could then adjust itself so that its solar arrays were facing the Sun and its batteries could begin to charge. But it wasn't working. The system couldn't make sense of the stars. And so *Odie* was now tumbling into space.

Was the team witnessing a repeat of what had doomed *Peregrine*? They couldn't be sure.

Star tracker relied on configurations of stellar patterns to determine the spacecraft's attitude. If there were errors in how that information was processed, however, or even the slightest errors in the database the system used to match what the cameras were seeing, the system simply couldn't identify what it was seeing. And if it couldn't do that, it couldn't orient itself.

As it turns out, the star tracker was indeed reading that stellar data, but was simply rejecting it due to an overly small margin of error built into the software. Meanwhile, the craft's helium thrusters were firing to correct the craft's attitude. But without a basis from which to align *Odie,* they couldn't stabilize the craft. *Odie* was somersaulting, and its batteries were draining fast. After years of work, and all the fanfare of the week, the mission was now in grave danger only moments after launch.

And yet there were also clues, if only they could react fast enough.

Occasionally, for instance, *Odie*'s computers would read a flash of power, drawing solar energy for the briefest of moments as the craft tumbled. It wasn't enough to charge the batteries, let alone sustain the mission. But it did offer hints as to the location of the Sun. Tim looked at the screens. If the team could isolate the precise moment when *Odie* signaled power, effectively taking a screen grab of the telemetry data that flashed across consoles, they could then use that data to find the Sun and manually override the trackers to orient the craft. It didn't have to be perfect. Just enough so that the array could catch a consistent dose of power from the Sun to keep it alive. Without charging, the onboard computer would soon die, along with any hopes of reaching the Moon.

All the team could do was wait for another glimpse of power.

They waited.

A flash of power registered across the screens.

"Now!" Tim bellowed.

Screen grab.

It wasn't precise. But it was enough of a clue as to the Sun's position. With it, the team could lock in the data, override the trackers, and halt the roll. It had worked. The solar panels were not quite Sun facing. But they were angled close enough to draw a charge. And that bought enough time "to think through the star tracker problem," Steve recalled.

Now charging, *Odie* was on its way to the Moon. The first major problem had been narrowly averted.

And yet there were more on the horizon.

· CHAPTER 23 ·

# LUNAR PAYLOAD

*Odie* wasn't exactly alone.

Riding aboard IM-1 was indeed a diverse mix of payloads, which included prototype navigation lasers, a student-built camera to film the landing, and a thermal cover developed by the Columbia Sportswear Company. But among those instruments and experimental tech were also the seeds of something that was perhaps far more ambitious: the makings of a lunar data center.

Just prior to *Odie*'s launch, I had lunch with the CEO and founder behind it, Chris Stott, whose company Lonestar Data Holdings sought to make history. We met inside a surfers' dive in Cocoa Beach: a smoothie-meets-kale kind of place where the menu options feature dishes called Feel Your Oats and Da Bomb. Chris's payload—effectively a software application—occupied one of *Odie*'s six commercial ride-alongs, which he hoped would serve as the nascent beginnings of a means of storing and processing data on the Moon.

"Our ultimate goal is global backup, refresh, and restore," Chris explained over breakfast. A sustainable lunar habitat would inevitably need on-site data processing and storage, helping with everything from managing life support, communications, and autonomous systems to reducing latency and reliance on Earth-based servers.

But Chris also envisioned the Moon as a kind of off-world backup for Earth's data—a hedge against catastrophic information loss of the sort that befell the Library of Alexandria, considered among the greatest archives of ancient knowledge, reportedly burned to the ground during Julius Caesar's siege in 48 BC.

The Moon, he explained, could thus house humanity's troves of digital data in relative isolation, much as the Seed Vault in the Norwegian archipelago of Svalbard serves as backup for the world's food crops. He wasn't alone in that thinking. In fact, researchers from Harvard and the University of Wisconsin–Madison concluded that the "risk of a catastrophic or existential disaster for our civilization is increasing this century," and that a storage system on the Moon could be a place to "preserve valuable information about the achievements of our civilization."

Of course, data was needed to operate, build, and scale sustainably. To do that effectively, it was advantageous to do computations, data processing, and storage locally.

"As you go further away from Earth, we can't always rely on our capabilities here," Gabrielle Hedrick, an aerospace engineer at MITRE, explained. "One of the big challenges is, 'How do you handle data?' "

Without local processing and analysis, bottlenecks were sure to emerge. In 2018, for instance, the Mars *Curiosity* rover team bemoaned its capacity to address a series of technical difficulties "because the amount of data coming down is limited." *Curiosity*'s project scientist Ashwin Vasavada wrote in a mission update, "It might take some time for the engineering team to diagnose the problem."

"Any expansion into space, for whatever reason, requires some sort of computation," Avi Shabtai, CEO of Ramon.Space, an Israeli-founded potential competitor to Lonestar, told me over the phone. "If you've seen images arriving from the Moon, it takes a lot of time," he explained. "The quality is not great, and all the processing needs to be done on Earth." But, he added, "if you can send it down to a data center on the Moon, run analytics on it, monitor it, save it, and then send results or conclusions to Earth, then it starts to make [business] sense."

Meanwhile, for Chris, the IM-1 mission was a proof of concept, just as concerns over escalating data energy consumption on Earth were mounting. Considered some of the most electricity-intensive buildings on the planet, data centers use up to an estimated fifty times the energy per floor space of the average office building. Growing consumer demands, as well as hefty new requirements for AI servers, were expected to balloon U.S. data energy consumption to 35 gigawatts by the end of the decade—more than double the amount consumed in 2022. To keep scaling, and keep pace with the coming tsunami of processing needs that AI- and quantum-driven technologies would demand, the network needed a lot more power. By early 2025, Eric Schmidt, former CEO of Google, who had recently purchased a controlling stake in the Long Beach–based Relativity Space, was sounding the alarm about the need for more power as the United States and China vied for dominance in artificial intelligence.

"People are planning 10 gigawatt data centers," he told House lawmakers about the future of AI and U.S. competitiveness. Schmidt, a Silicon Valley veteran who had played a major role in scaling the last tech wave with the internet, now seemed to be placing his bets on the next one relying on space, where a natural cooling environment and constant solar power could begin to offset the kinds of yawning energy, water, and land needs that AI-fueled systems would demand.

"Gives you a sense of how big this crisis is. Many people think that the energy demand for our industry will go from 3 percent to 99 percent of total generation. One of the estimates that I think is most likely is that data centers will require an additional 29 gigawatts of power by 2027, and 67 more gigawatts by 2030. These things are industrial at a scale that I have never seen in my life." Journalist Eric Berger later opined on social media that Schmidt had acquired Relativity "as a means to support the development of data centers in space. Such data centers, ideally, would be powered by solar panels and be able to radiate heat into the vacuum of space."

"This probably helps explain why Schmidt bought Relativity Space," he posted on X.

The next day, Schmidt dispelled any further speculation with this simple reply: "Yes," he posted. Emerging notions of space-based data infrastructure suddenly seemed all the more real.

China, however, was already working on it; its Three-Body Constellation project potentially marking the dawn of the most powerful, distributed AI and space-based computing system ever deployed. To afford sufficient power, as mentioned, Beijing had plans to build a one-kilometer-wide solar-powered array. That collected energy could then be converted into microwaves or lasers and delivered wirelessly in space, or to ground stations on Earth, where it could be converted into electricity for use.

"It is as significant as moving the Three Gorges Dam to a geostationary orbit 36,000 km above the Earth," said Long Lehao, a scientist and member of the Chinese Academy of Engineering. "This is an incredible project to look forward to."

Of course, it wouldn't be easy. For starters, computer chips generally do not do well in space. As spaceships venture beyond the protection of Earth's magnetic field, high-energy radiation bombards microprocessors and damages chips in ways that can lead to operational failure. Russia's space program experienced that the hard way in 2011, when its Phobos-Grunt, an interplanetary probe designed to land on the Martian moon Phobos, crashed back down to Earth after cosmic rays were suspected of frying the craft's SRAM chips soon after launch, even though subsequent investigations pointed to programming and engineering issues. New designs and polymeric matrices could make those systems more dense and robust. The problem was that some of those higher-energy rays were more likely to interact with dense material, which created a shower effect of secondary particles that degraded systems, and ultimately underscored the need for development of new materials that could better absorb radiation, or heal themselves.

Shielding, however, can also take other forms.

Lunar lava tubes, for instance, appeared effective at blocking much of the Sun's radiation. Yet over time, those habitats would require more than just solar power. To solve for that, in April 2025, Beijing and Moscow announced a bold plan: The two nations would endeavor to build the Moon's first nuclear power plant. Slated for completion by 2036, the facility would use a Russian reactor capable of powering a sprawling, Disneyland-sized base constructed across the Moon's surface. But that wasn't all. Those plans also included a lander, hopper, orbiter, and rover—along with a Skynet-like satellite constellation to surveil the Moon and oversee assets and monitor rivals in real time. Considered among the Earth's largest video surveillance network, with more than 600 million cameras, Skynet, or Tianwang, is where the upcoming "optical surveillance system for the Lunar Research Station can draw on the successful experience," China's main space agency published in the Chinese academic journal *Acta Optica Sinica* in 2024. At its core was an explosion of data, emanating from virtually every system, which demanded new levels of power, processing, cooling, and storage.

Precisely how those systems might be devised now seemed to be the question of the moment. While Chinese authorities were long experienced in devising powerful, centralized architectures, America's relative advantage seemed to lie more in the fostering of commercial markets. Retired U.S. Air Force lieutenant general Steve Kwast perhaps put it best. "We can't beat communists by using big-government methods," he wrote.

"China is much better at central planning than America will ever be. China is dominated by the Chinese Communist Party (CCP), a unitary dictatorship that controls all the levers of power. The CCP is faster and more efficient than America will ever be using the tools of socialism," he added. "We need to play to our unique strengths. America's distinct competitive advantage is our culture of freedom, which has created the world's leading environment for entrepreneurial innovation."

That was just the sort of sentiment Chris, Steve, Kam, and Tim, and the countless other new space entrepreneurs, wanted to hear. Still, Chris needed someone who knew how to harden computer systems not just for space but for the lunar environment in particular. It was a short list. And a man who I'd run into just outside a hot chicken joint in Nashville, Tennessee, was on it.

Dennis Wingo was the founder turned chief technology officer of Skycorp, which had built the RISC-V-based web server for the ISS. He was figuring out the spice level of his order when I said hello.

"It's good to see you in person," I exclaimed, having only spoken with the man a few times over the phone. We happened to be rerouted through the same airport around the holidays, even though Dennis clearly had more pressing matters on his mind. He was being asked by Chris to sort out data infrastructure for the Moon, where lunar dust and drastic temperature fluctuations can cause molecules within the electronics to radically (and destructively) expand and contract. And yet for Lonestar, Intuitive Machines, NASA, or their Chinese competitors to truly make use of space and the Moon, it seemed they'd need to not only rekindle the spirit of those first explorers, but also utilize a kind of fusion of technologies as had been done during the age of Apollo.

· CHAPTER 24 ·

# HUNTSVILLE

At first glance, the old seat of Alabama's state government might have seemed an unlikely birthplace of America's leap into space. But by the early 1960s, that transformation was under way, a product of technical expertise already present with rocket weaponry developed at the Redstone Arsenal, just south of the city, as well as a few well-placed political favors, born of an Alabama congressional delegation keen on bringing new investment to the region. Nestled in the Appalachian foothills, Huntsville would become home to Marshall Space Flight Center, tasked with developing the Saturn V rocket, eventually overseen by Wernher von Braun, a German-born aerospace engineer.

While von Braun would become a household name, other German scientists like Ernst Stuhlinger, whom Dennis Wingo would meet years later while doing research at the University of Alabama in Huntsville, would become instrumental in developing the kinds of propulsion needs that would arise after the Apollo program. Stuhlinger, in fact, had been focused on a system that expelled ions and electrons to generate exceptionally high exhaust velocities, just the sorts of foundational concepts that would be integrated into the systems that powered NASA's Deep Space 1 mission, as well as the agency's Dawn mission, which explored the asteroid belt between Mars and Jupiter, and that is also now widely employed among commercial satellites across the space landscape.

But Stuhlinger also had bigger ideas.

In a seminal 1955 paper titled "Electrical Propulsion System for

Space Ships with Nuclear Power Source," he even proposed the development of an ion-propelled spacecraft that would employ a nuclear reactor to heat up fluid that would power a turbine, which would in turn provide the electricity needed to accelerate ions as a propellant at speeds ten times that of chemical engines. Not only would the craft make lunar, and perhaps even interplanetary travel more feasible, but conceptual designs for the thrusters themselves were thought to showcase continuous thrust at max power for years at a time.

The concept, like much of technology itself, built on earlier ideas. In this case, the beginnings of conceptual plans using electrostatic forces for propulsion had shown up as early as 1906, in Robert Goddard's notebooks, as well as in the writings of Hermann Oberth and Konstantin Tsiolkovsky, considered early pioneers of astronautics and human spaceflight. But it was Stuhlinger who advanced it. In one instance, he envisioned the system to be used in the design of a kind of lunar ferry, the largest of which weighing approximately 78 tons—only slightly smaller than today's Starship upper stage (without propellant)—could be used to supply and build what was now in development: a Moon base. He envisioned a much larger craft (roughly 280 tons) would be needed for a Mars mission.

Of course, this was a different time. Just two decades had elapsed since the world had entered the atomic age, and visions for space had grown correspondingly bold. Many were built on the early work of the more than sixteen hundred Nazi scientists and engineers, often referred to as "Our Germans," who had been scooped up in a secret U.S. intelligence program toward the end of World War II. Known as Operation Paperclip, the initiative had harnessed German know-how to advance America's space program, and to ensure it was not co-opted by the Soviets, who would ultimately corral more than twenty-two hundred Nazi scientists to work on advanced rocket designs in facilities just outside Moscow.

Before the war ended, Stuhlinger had worked in the little north German village of Peenemünde, which von Braun had chosen as a secret locale to develop and test the very V-2 rockets that rained down on London. Many scientists and engineers had left at the war's conclusion, complicating U.S. and Soviet intelligence efforts to track them down. Once they did, however, the terms were often stark. A 1989 interview with Stuhlinger, in fact, revealed the nature of just how he had ended up in Alabama:

> In October 1945, I was contacted by an officer from the United States, who asked me if I would like to accept a contract to come over to this country and continue work on rockets. I asked him first with whom I would work and he said, "Well, your boss Werner [*sic*] von Braun is just about to go over there." Then I asked him what we would be doing. He had a very short answer. He said "you will be doing what the United States of America thinks is best." But he was quite reassuring. He said, "I bet you will have an interesting life."

From there, Stuhlinger would toil on guidance and navigation instruments for the Explorer 1 project and Saturn V, before helping design the solar X-ray telescope used in America's first space station, eventually becoming director of science at the NASA Marshall Space Flight Center. But in those early days of the first space race, he recalled stories with a hint of excitement, referring to Sputnik as a "wonderful wake-up call for us Americans" and saying that he "immediately felt a kind of thankfulness to the Russian colleagues."

And yet to compete, there also seemed to be a growing recognition that the emerging nexus of aerospace and computer engineering would have to be better leveraged. These fields would have to be strategically integrated, which meant embracing a new and—in many ways—commercially driven wave of revolutionizing technologies.

Of all the technological triumphs of the era, few would prove more influential than the transistor. Used to amplify or switch electronic signals, transistors were a vast improvement over bulky, inefficient vacuum tube technology, serving as a kind of gate for electrical current that allowed signals to be manipulated with greater precision. That, in turn, rendered more compact and complex circuitry, thus opening the door to a host of new consumer products, including transistor radios, pocket calculators, and videocassette recorders, gaining traction as the industry continued to push the limits of miniaturization and efficiency.

In that furnace of innovation and commercial appetite, the microchip was born. NASA and the Pentagon became early adopters, purchasing them in bulk, and thus ensuring a nascent market that would become a cornerstone of America's burgeoning tech industry. The Apollo Guidance Computer, which was used aboard both the command module and the lunar modules, for instance, employed some five thousand of these integrated circuits (or microchips), just as manufacturers like Fairchild Semiconductor and Texas Instruments sought to fulfill new and growing demands. The technology offered a range of new options when it came to processing power that benefited navigation and guidance, automatic course corrections, and a host of other novelties. The system could handle eighty-five thousand instructions per second, allowing for real-time corrections and three-dimensional adjustments. Impressive computational power at the time, until one considers an iPhone 15 Pro, which can perform more than two teraflops—or roughly two trillion floating-point operations—per second.

Back then, however, that kind of computational power with hardware small enough and light enough to fit inside a payload fairing was virtually unheard of. Meanwhile, early microchip pioneers like Gordon Moore and Robert Noyce decided to leave Fairchild Semiconductor to start a company called Integrated Electronics one year before Apollo 11 touched down on the Moon, thus becoming among the germinating seeds of America's tech hub in Silicon Valley.

Today, it is better known as Intel.

The personal computing age had begun, with the Kenbak-1 (a machine equipped with 256 bytes of memory and a single circuit board) hitting the market just three years later in 1971. Three years after that, it was the Altair 8800, which lacked a keyboard but nonetheless graced the cover of a 1975 edition of *Popular Electronics,* where it was billed as the "World's First Minicomputer Kit to Rival Commercial Models." It sold for $395. Thousands lined up to get one, as programming language grew more sophisticated, including something called C programming created by scientists at Bell Labs, which was used for the still-developing Unix operating systems that allowed for more real-time responses in computing power—a function that was beginning to see new applications in the worlds of both automotive and aerospace design.

By 1976, Steve Jobs and Steve Wozniak had unveiled the first Apple computer, after building it in the garage of Jobs's parents' house in Palo Alto, California. The Apple 1, as it came to be known, included a single motherboard, came preassembled, and seemed to feed the appetites of hobbyists and computer enthusiasts, many of whom also seemed quite enamored with the feats of the Apollo program. Coincidentally, it was within this climate of accelerating innovation that Dennis Wingo came of age.

"I knew I had to be a part of this," he told me.

Having grown up just over an hour's drive from Huntsville, Dennis had developed both a knack for computers and an interest in space. The pace of development between the two, however, appeared uneven, disparities in pacing becoming ever more evident.

"I could see through that whole time that the aerospace industry and the commercial computer industry diverged, and the commercial computer industry was making advances way faster than NASA and defense," he said. "I wanted to bring some of those advances into NASA and defense industry. And that's the last thing in the world they wanted."

That, of course, was then.

The paradigm now appeared to be changing. SpaceX, Blue Origin, and a torrent of start-ups—many of which were led by tech industry veterans more accustomed to the fail-fast ecosystems of Silicon Valley—were now competing against legacy firms, just as traditional barriers to space were falling.

"We're going to live in a different world in forty years in space," Dennis told me. "It's as big a revolution as the computer industry has been in for the last forty years, and I've been a big part of it for the last forty years."

This was the moment for which he had been waiting: a long-anticipated convergence between aerospace and those rapidly advancing frontiers of computing. Working alongside Chris Stott, he was on the verge of something extraordinary. If all went well, Intuitive Machines' next mission to the lunar south pole would carry with it a small data center.

IM-1 would be a proving ground.

· CHAPTER 25 ·

# EN ROUTE TO THE MOON

By 7:30 p.m. EST, Steve Altemus was in a T-shirt. He looked ragged. They all did. It was day three (and fourteen hours) into the IM-1 mission. And a quick walk through the corridors outside mission control revealed the remnants of what looked like a slumber party. Fold-out cots and unkempt blankets were jammed beside office

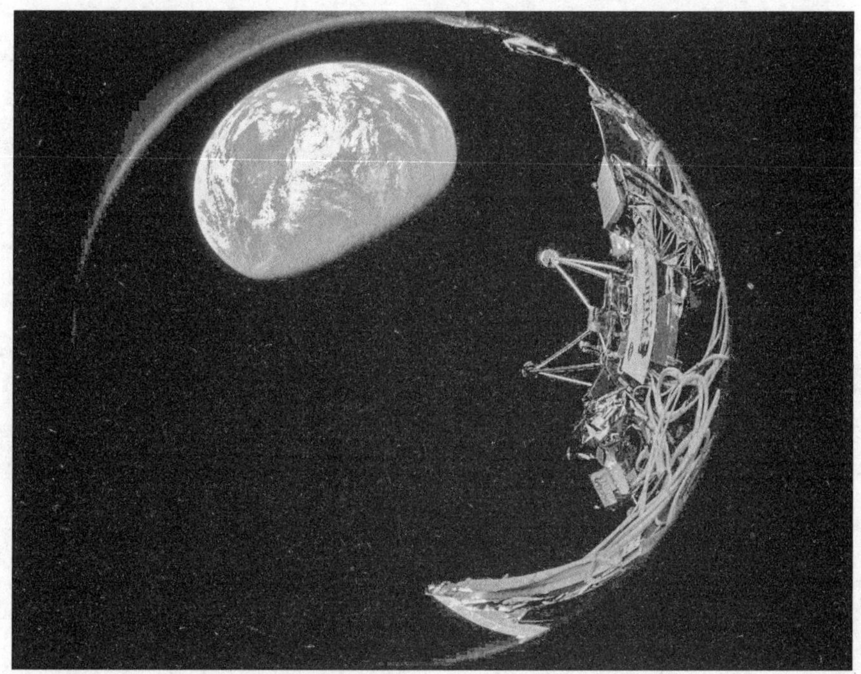

*Odie* traveling to the Moon, February 2024.

chairs. The scatterings of delivery orders and water bottles filled the break room.

It was February 18, 2024, and *Odie* was now in cislunar space, a loosely defined region that stretches from Earth's geosynchronous orbit to the Moon. The craft would soon complete its second engine burn on the way to lunar orbit. Inside mission control, meanwhile, eight people, including Rob and Tim, were monitoring telemetry data as it came in. Three teams of eight (code-named red, white, and blue) were each pulling eight-hour shifts around the clock. Steve oversaw a fourth team, the planning group. And by now, *Odie* was prepping for lunar orbital insertion, which would require flying over the landing site and swinging around the back side of the Moon. That would cause a signal loss, given the Moon blocks radio signals. Everyone back in Houston would just have to wait. In 1969, moments before that Apollo 11 communications blackout, Bruce McCandless famously radioed ahead to say the craft's "systems are looking good going around the corner."

"We'll see you on the other side, over," as the command and service module dipped out of comms. From back there, the cosmos suddenly seems "awash with stars," as Al Worden, command module pilot for Apollo 15, later described it. Devoid of Earth's light pollution and atmosphere that would otherwise obscure the sheer luminescence of the Milky Way galaxy, the far side of the Moon offers unparalleled views. Its extreme blackness is a comparably ideal place for stargazing and radio astronomy—something that had long excited both American stargazers like Steve Durst and Chinese ones like Xue Suijian. Distant galaxies and colorful nebulae reveal themselves in ways usually not detectable from Earth.

"TIG minus three minutes," Tim called out, a reference to the time of ignition, or engine burn, where *Odie*'s thruster would correct attitude and increase speed. The craft was about half the distance to the Moon, with another 126,000 miles to go. An upcoming burn would pick up the pace. By now, the craft's tanks were pressurizing, which maintains the integrity of the fuel lines and ensures that fuel is being pushed out of the tanks and into the engines. Unlike on Earth, where gravity helps in that process, a lack of pressurization in the vacuum of space could allow the remaining fuel to float away from the injector. Firing the engines in that event could lead to leaks or ruptures in the main line, and even explosions. Virtually every mechanical assumption on Earth had to be reconsidered for space.

"Sixty seconds."

The methane fuel line prep wasn't complete. Cryogenic chilling of those lines was key to preventing pressure buildup, which could cause leaks or ruptures and thus jeopardize the mission.

"Five seconds to TIG," Tim shouted out. But readings on the fuel lines didn't look right. Even if it was properly cooling, the orientation of the craft might have exposed the fuel line to more Sun than expected, which would then require more cryogen to keep temperatures nominal.

"I don't think this is going to work," Rob chimed in.

"Okay. . . . Abort."

They would try again in a few hours. But it wasn't great news. The initial window had been missed. And there would be consequences. *Odie* had expended precious helium to correct attitude for a burn that never came, having already done so in its initial flurry of adjustments to point its solar panels toward the Sun. With razor-thin margins, too many adjustments could make those helium tanks run dry before *Odie* would be able to reach the Moon.

What's more, as the craft got closer to the Moon, something more troubling was happening. Twice, the automated guidance software had misfired the helium thrusters, venting precious propellant. In doing so, *Odie* had lost a staggering 21 percent of its total helium supply. It was a blow that had serious implications for the mission, especially given that the last trajectory correction maneuver and burn were coming up. And yet with so much helium lost, a wave of alarm swept through mission control, especially because the final leg of this lunar voyage demanded the kind of precision and control that helium made possible. It was critical for pressurizing the craft's propulsion system needed to effectively pump the brakes during lunar descent, bringing a craft traveling at thousands of miles per hour to a mere walking pace, before they could hope to stick the landing. But before that could happen, the timing of *Odie*'s deorbit also had to be exact. If they missed the landing site, another lunar orbit would risk running out of propellant altogether.

And yet fatigue was starting to show. Many had been working on only a few hours of sleep over the last few days. The bank of screens glowed against a collection of faces that now revealed the beginnings of dark circles under several pairs of eyes. Steve looked especially run down. Rob was still in mission control, as were Tim and the rest of the team. Some of them had been going almost nonstop for nearly forty-eight hours.

"We're twenty minutes to TIG," Tim called out. "This is the time to enhance your focus. We're at the end of a long shift."

By now, *Odie* had only 112,000 miles left in its journey. The Moon's gravity would soon start to influence the direction of the craft, which meant *Odie* would have to expend still more fuel to counteract those effects and maintain trajectory. If left purely to the Moon's gravity, the craft could crash into the lunar surface or fly right by it.

"Prep oxygen complete," Tim shouted. Just before the engines lit, the room seemed to fall quiet.

"Burn in progress," Rob said, breathing out.

"Loss of signal," chimed in the avionics engineer Jordan Reynolds, noting another communications blackout. Again, this part had been expected. The second burn repositioned *Odie* so that its tail faced Earth, causing the craft's antenna to point away from ground stations. But, provided the burn went well, the attitude of the craft would soon correct, and communications could be reestablished. The time read 8:36 p.m. Rob was sitting at his station, hands clasped, occasionally rubbing his forehead. Eight minutes had elapsed with no signal. And still nothing.

*Odie* was not responding.

"Take stock of your systems," Tim called out. "Number one priority is making sure the vehicle is okay." By now, they could only wait, which had another, perhaps unexpected, effect. As Rob waited for a response, he couldn't help but think of those darker days two decades earlier, knowing the silence he and flight controllers suffered after the cascade of problems that had doomed the *Columbia* shuttle. Today, it was a different mission control. But he was still in Houston and again waiting for a signal.

Then it happened. A blip on the screen.

"We've got a safe vehicle!"

*Odie* was alive. The engine burn, which had fired for just eight seconds, was good enough to avoid a third trajectory-correction maneuver. That part was critical. Now the team needed to preserve what was left of *Odie*'s propellant. The vehicle was just about forty thousand miles from the Moon. With that, Trent Martin, the company's vice president of space systems, took the comm.

"We are good to start transition from blue to red," he said, referring to the next eight-hour shift. It was time for the blue team to rest. A series of low-slung shoulders paraded out into the hallway. Rob, however, had moved inside one of the corridors' side offices, where a separate team sat in front of a bank of computers. "That guy isn't going anywhere," explained software release engineer Travis Roberts. It was Travis's job to manage much of the software used in *Odie*'s systems, including guidance, navigation, and data processing.

"Please text me as soon as you get that data," Rob said.

He then noticed me standing there.

"Good to see you," he said, extending his hand and exhaling as he did. Rob then leaned against a glass door. "Not hearing from *Odie* for twelve minutes . . ." He breathed out again, not finishing the sentence.

"Get some rest," I replied.

He smiled and then vanished into an adjacent office for yet another impromptu meeting.

· CHAPTER 26 ·

# LUNAR DESCENT

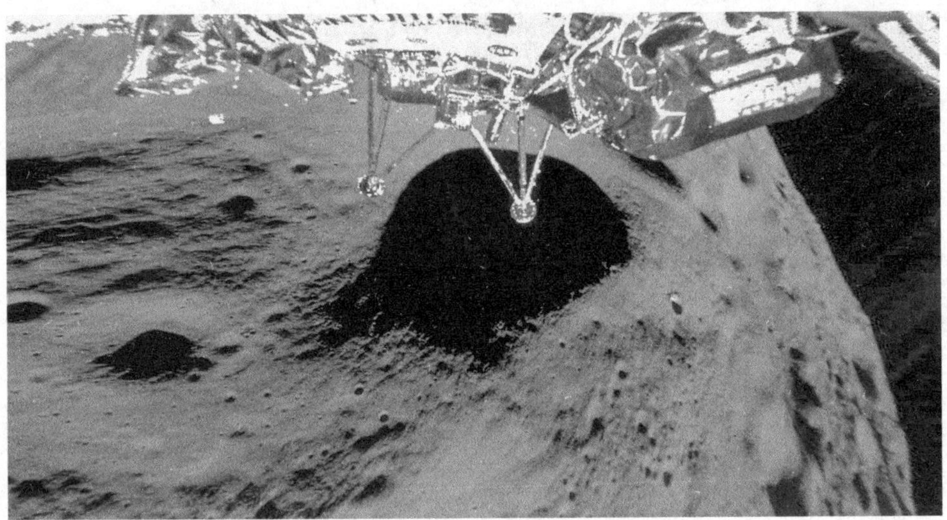

IM-1 lander prepares to touch down near the lunar south pole, February 22, 2024.

*Odie* was now approaching the Moon. What had been a distant glow from that Cocoa Beach barbecue now revealed to the vehicle's cameras a pockmarked landscape. The team would ultimately focus on Malapert A, a forty-three-mile impact crater near the south pole, which was a candidate for the future Artemis lunar base camp. But from the imagery generated by NASA's Lunar Reconnaissance Orbiter, it didn't look like an easy place to land—deep valleys of perpetual darkness and soaring mountains, which included that towering Malapert massif itself. The ground below could be unstable given the greater depth of the regolith, which is basically just rock chips and a glass-like dust. India's Chandrayaan-3 (not a commercial craft) had landed relatively close to there, though still more than 370 miles from the Moon's south pole. *Odie* was shooting for less than half that distance, and the vehicle had just entered lunar orbit, having completed its scheduled 408-second main engine insertion burn.

It was now circling the Moon, but was extremely low on helium. All those adjustments during flight had taken their toll. If the helium ran dry, *Odie* would risk coming in too fast. Rob had been running the numbers from the previous thirty-six hours. It seemed as if the mission would run out when *Odie* needed it most—just as the craft endeavored to land on the Moon. Meanwhile, as engineers like Rob scrambled to assess the problem, another problem began to stir, one that could prove catastrophic. Somewhere in that stream of telemetry data, the first signs were emerging.

The craft had dropped in over the lunar north pole, and had been in a far more elliptical orbit than expected. The last engine burn had gone on a bit too long. The propulsion exhaust after the main engine cut off had given *Odie* an extra bit of oomph that put the craft in a slightly oval-shaped path, rather than a circular orbit. It was a correctable issue. But as the team evaluated their options, a new wave of data sent a chill through mission control.

*Odie* appeared to be flying dangerously lower than expected.

A lot lower.

Tim instructed the craft to fire up its laser range finder to verify the reading, bouncing pulses off the surface and then measuring the time it took for the signals to return. The altimeter had a range of about fifty miles. *Odie* should have been flying roughly sixty-two miles above the surface. If the device couldn't get a reading, it meant one of two things: Either *Odie*'s altitude was perfectly fine, and was just flying higher than that fifty-mile threshold, or the altimeter wasn't functioning at all—a more terrifying prospect—which meant they'd have no real sense of how high above the lunar surface *Odie* was actually flying.

There was no response. They waited. Still nothing.

Steve didn't trust that lack of a reading. The company's communications and trajectory teams were working to get answers, including calling the very contractors who had built the device. But they now had conflicting information from other instrumentation. They needed to understand why. More calls. More investigating. The pieces were coming together.

It wasn't a software glitch.

It wasn't a hardware failure.

The culprit was far more maddening. The safety catch. The fail-safe mechanism designed to prevent catastrophe was, in a cruel twist, causing what could now be a mission-ending problem. *Odie,* in other words, couldn't "see" the lunar ground as it came in for a landing. The sensor that was crucial for gauging altitude had never been fully armed. It turns out, it had just not been disengaged prior to launch. The device was not considered "eye safe," meaning its laser emits at wavelengths potentially harmful to the human eye. So it requires a safety catch, which had to be disengaged. It was purely a manual function. If it wasn't flipped, there was nothing that could be done now. The team had run thousands of safety checks prior to liftoff. But they were exhausted. And it was missed. Pure operator error. Without an altimeter to guide them, the team had no conception of how extremely low *Odie* was now cruising above the surface of the Moon.

As the craft orbited, star tracker data would eventually paint a harrowing picture: The vehicle had actually dipped below the lunar horizon. That would normally mean a crash, putting it on a trajectory below the lunar surface. Yet in a twist that defied probabilities, it seemed *Odie* had gotten lucky.

Shortly after lunar orbital insertion, it had entered what appeared to be a lunar valley, whose walls cradled the spacecraft and bought mission controllers the precious moments they needed. Once they realized how low *Odie* was flying, they instructed it to climb, and fast. *Odie* began ascending. It was just enough. Had it not been for that valley, *Odie* would surely have ended up a wreck like so many of its predecessors. The lander had been spared, at least for the moment. But the relief would be short-lived. There was still the issue of the altimeter.

They'd have to land on the Moon without the very instrument designed to get them there.

"Tim," Steve said. "We're going to have to land without a laser range finder."

Tim's face drained of any color. How could they land on the Moon without an altimeter? *Odie* suddenly seemed destined for the same lunar scrap heap where the majority of Moon missions had ended. Could something be done? Could they improvise? Could they somehow hack the system to trick it into overriding the safety feature? They'd have to run simulations. And there just wasn't time. There were too many variables.

Tim and Steve left the console together. They were now walking in one of the office's main corridors. *Odie* was facing imminent failure. But other than the small team in mission control, almost no one else knew it. Meanwhile, at the company's operations center, a catered party of esteemed guests—including Johnson Space Center director, Vanessa Wyche; Apollo 11 flight controller Bill Moon; Ax-2 pilot John Shoffner; and Doug Terrier, Johnson Space Center's associate director for vision and strategy—were all waiting for news. Intuitive Machines had been broadcasting many of its activities on NASA's live feed. Partygoers mingled and made nervous small talk as the clock ticked closer to the expected landing time.

Then, back in mission control, Tim had an idea.

Included among *Odie*'s six NASA payloads was an experimental navigation system, which used three laser beams to measure the lander's velocity and altitude relative to the lunar surface. It was supposed to be a tech demonstration, merely gathering data during lunar descent, perhaps to be one day employed during Artemis Moon landings. But it was purely experimental, employing a Doppler effect by analyzing laser light reflected off the Moon.

It was never meant to be relied upon so soon.

But what if, Tim pondered, they could use it now to save *Odie*? The device would effectively have to be hot-wired through a software rewrite to guide landing, which would normally take weeks of testing. They had two hours.

Out in the observation rooms, the officially scheduled landing time switched from 4:24 ET to 6:24 ET. Chatter at the party grew more nervous, and the company's stock plunged 8 percent in after-hours trading.

This was it. The team was out of options. *Odie* could make one more sweep, losing communications with mission control with its signal cut off as it traveled around the lunar far side. Then it would begin its descent. Meanwhile, software teams would have to run a series of simulations and devise a patch before *Odie* circled back. By then, fix or no fix, the vehicle would have to land. There would be no second chance. Any longer, regardless of altimetry, and it would be out of helium. To get the software right, however, the sequencing would have to be precise. Reminiscent, perhaps, of those yesteryear efforts that saved the Apollo 13 module, with the craft's sequencing turned on in a particular order, *Odie*'s systems "did not like being rebooted like that." The wrong sequence could create a host of new problems, including not turning on at all or sending the craft spiraling out into space. Plus, once lunar descent began, there would be no going back.

Software teams were now running simulations. *Odie*'s systems had been duplicated at IM headquarters in Houston so that they could test out just how the hardware might react before uploading the patch. As the craft swooped back around, controllers in Houston uploaded what they had to the ship's main computer.

Thirteen minutes to touchdown.

It was 6:11 p.m., and the craft's helium levels were showing critical. Rob had been running scenarios. But with no new data coming in, there was only so much he could do. Now, with *Odie*'s signal having returned, he "could see the engine starting to depressurize and the thrust level going down." This was the helium problem they had worried about.

And *Odie* was now becoming more unstable. They hoped that by throttling down, it would still have enough to make it to the surface. Hundreds of procedures had to be executed for a safe landing, much of it depending on the craft being able to read data regarding altitude and horizontal movement. But once again, communications had halted. Signal loss was expected during descent, given the engine creates a cone of electromagnetic interference during landing. As the minutes ticked by, an uneasy silence washed over both mission control and nearby observation rooms. It was almost as if the entire building, and indeed much of the viewing global audience, were now collectively holding its breath. And for the moment, there was just no way of knowing whether the jerry-rigged altimeter had worked. Steve stood at the center console with arms folded. The blue team focused on their monitors.

*Odie* was on autopilot.

Two minutes to scheduled touchdown. The craft's maneuvers were on timers. The craft should have been pitching and getting closer to the surface. But by now, those two minutes had come and gone. Then Tim's voice broke through over comms, ringing out across mission control and the adjacent operations room.

"We're not dead yet."

A few encouraging claps ensued. But still, everyone waited. Steve stood at the center console. Rob was hardly moving. Twelve team members were now in mission control, all looking for signs of life. There was nothing . . . until there was.

A faint signal.

Tim took the comm.

"All stations, this is mission director IM-1. We're evaluating how we can refine that signal, dial in the pointing for our dishes," he said. For a moment, the room held still.

"But what we can confirm, without a doubt, is our equipment is on the surface of the Moon. And we are transmitting."

With that, catharsis. An explosion of celebration ensued, like a dam breaking with a rush of hugs and high fives. Flutes of champagne were raised. And Steve then took the comm.

"I know this was a nail-biter, but we are on the surface, and we are transmitting," he said. "Welcome to the Moon."

The hot wire had worked, although not perfectly. It turns out, as *Odie* descended, the craft had been traveling roughly three times faster than anticipated. It also had a lateral velocity of about two meters per second, akin to a jogger's pace. The landing site was also slightly off. Rather than a flat-surface touchdown, *Odie* had mis-

IM-1 lander having touched down near the lunar south pole.

Steve Altemus (*in suit, right*) celebrating following IM-1 landing at Intuitive Machines' headquarters in Houston, Texas, February 2024.

takenly endeavored to land on a lunar slope. And as it descended, it skipped off the surface at least three times, apparently snapping off at least one of its landing legs. Without that leg, it tipped, eventually falling over. Some antennas now faced down toward the lunar surface. The team would have to reroute communications, with just a single solar panel now facing the Sun. But it was enough to supply the lander with the electricity it needed to complete much of the mission, roughly 170 watts of power. *Odie* was on its side. But it had landed, and was now functioning. Moments later, NASA's administrator, Bill Nelson, beamed his congratulations over webcast. "Today for the first time in more than a half century," he said, "the U.S. has returned to the Moon."

· CHAPTER 27 ·

# THE SPACEX EFFECT

History had been made. The IM lander, despite an imperfect touchdown, had become the first commercial craft to reach the lunar surface, marking the dawn of a new era in private space exploration amid intensifying competition with China. Yet even amid that applause in Houston, the cadence of commercial space pressed forward. Just days earlier, as that Falcon 9 rocket prepared to carry *Odie* into orbit, SpaceX had three other rockets standing on launchpads across the U.S.; each poised for entirely different missions. It was a vivid snapshot of a rapidly evolving launch industry and a reminder of just how firmly SpaceX was positioned at its center. One was at Vandenberg Space Force Base in California. Another was preparing for its third mission of the year under a National Security Space Launch contract. And Starship was gearing up for a launch from Starbase in Texas.

It was a remarkable pace, even for a company that had tripled its launch cadence in the previous two years, outpacing Russia's and China's national programs combined and accounting for roughly 80 percent of all payloads delivered into orbit. SpaceX was the clear global leader. And its emphasis on vertical integration was central to that success, as were its reusable rockets—a concept whose origins had actually been derived decades earlier.

Back in the early 1990s, a subscale prototype of a single-stage-to-orbit vertical landing vehicle, built by McDonnell Douglas and known as the Delta Clipper, had in fact emerged as an early pioneer in reusable technology. It employed four RL-10A5 engines and four retractable landing struts, leveraging a variety of commercially available parts. Even though the program was ultimately discontinued, it's hard to imagine those base concepts did not at least indirectly influence a seminal staff meeting at SpaceX years later. Musk had gathered his team to showcase an animation of the kinds of vertical landings that he hoped to achieve.

A former air force colonel named Lee Rosen was in attendance, watching from Vandenberg Space Force Base in California, after having only recently joined the SpaceX team from U.S. Air Force Space Command, later redesignated as part of the U.S. Space Force under President Trump. Lee had first met Musk during a 2010 visit at the Kennedy Space Center in Cape Canaveral, where the SpaceX founder was scheduled to meet with President Obama. As the two men waited for the president to arrive, Elon "was asking all kinds of interesting questions," Lee recalled. "There was not much else to do while you're waiting for an hour or so for the president to show up. And so he started asking things like 'How would you land a rocket?' I'm like, 'Well, you don't land a rocket.' He said, 'No, no, I think you can. Do you know any other mode of transportation that you throw away after one usage?' I said, 'No, I really don't.' He said, 'Imagine how much your airline ticket would cost if you threw away the airplane every time you flew it.' "

As the two men stood waiting in that Florida hangar, Lee was intrigued.

"We talked about parachutes . . . landing gear and wings, and then settled on propulsion. Propulsive landing, where you turn the engines back on, would be a way to land a rocket. And I remember talking about parachutes and he said, 'Well, parachutes don't really work well on Mars.' And I'm like, 'Wow, this guy's kind of out there. But interesting.' And just during the course of the conversation, I started to think to myself, 'This guy could change the way the business is done.' "

The old model indeed was ripe for change.

"Prices had gone through the roof," Lee added. "[NASA's legacy contractors] were super reliable. And we were brought up in that industry to say that there is no cost too high for mission assurance, and that our nation's blood and treasure is on the line. All of that's true, but perhaps there would be a way to save some money and have high reliability."

Six months later, Lee decided he'd find out, and put in for his retirement to join SpaceX. The company was about to break ground at Vandenberg, taking over the launch (and soon to be landing) site at Space Launch Complex 4 East, which needed to be retrofitted to accommodate the Falcon Heavy, then considered among the most powerful launch vehicles since the Saturn V. Colonel Richard Boltz, who had just taken command of the Thirtieth Space Wing at Vandenberg, called it "the next chapter" in the military's relationship with the space industry. And Lee was now tasked with getting the facility ready. He was given a year to do it.

Fortunately, he had the freedom to pick his team, a moment that coincided with a floppy-haired former NCAA sailing champion who would soon come into Lee's orbit. Frank Tybor had grown up in Coronado, California, had a penchant for yacht racing, and had studied aeronautical and astronautical engineering at Stanford University. He had an airy, affable, easygoing charm. But there was also no mistaking his technical acumen, one that he would later put to use on Starship, a skycraper-sized rocket, and the most powerful launch vehicle ever built. The team was assembling. Others like Zach Dunn, who would go on to become COO at Relativity Space, and Ricky Lim, who would become senior director of launch operations at Cape Canaveral for SpaceX, would also join. The effort for something bigger seemed to be coming together, with a twofold emphasis: speed and cost. They were tasked with moving quickly and cheaply. Cost cutting had become a company hallmark. And that would extend to something as basic as laying down an access road.

"We had to build a road from this hangar to the launchpad, and it went around the side of a mountain," Lee explained. Eight hundred fifty feet long. At least seventy-five feet wide, reinforced with steel to account for the width and weight of the rocket and its transporter. The initial quote was $2 million.

"We negotiated with the contractor, and with doing some work ourselves, we got it down to a million," Lee explained, a figure he then proudly presented to Musk.

"He looked me right in the eyes and said, 'Have you lost your fucking mind? I just paved my driveway and it's longer than that. It was $200,000. If you can't do it for half a million, I'll find somebody who can,'" he recalled of Musk's response.

Lee went back to his team.

"And guess what? We ended up doing it for half a million dollars. A big cost of the road was the stuff that goes underneath. They call it road base. We needed a couple of feet thick of road base. And we had all this used concrete that was on-site that we had from the old Titan pad. So we rented a concrete crusher, and laid it all down ourselves."

"There's a single *s* between 'scrappy' and 'crappy,'" Frank added, smiling. "SpaceX prides itself on being on the right side [of that equation]."

Within sixteen months the launch facility at Vandenberg was complete.

"We didn't quite hit Elon's year mark, but we did it for way cheaper and way faster than I thought was humanly possible," Lee recalled. By 2015, he had become SpaceX vice president for mission and launch operations, and Frank had taken over as senior manager for recovery engineering and operations. The focus was now centered back on landing rockets.

"To bring a rocket back to land, you're reentering backward at about five thousand miles an hour and you have to get an engine to start-up burn and have a controlled descend," Frank explained, describing the retrograde burn that essentially slowed their fall.

"Flying backward at Mach 5.

"Imagine taking your boat full throttle at thirty-five knots on the lake," he added, "and then jamming it into full reverse and expecting the engine to actually do that without exploding." The more technical term for the maneuver is "supersonic retro-propulsion," effectively needed to achieve a soft landing on Earth—or the Moon or Mars, for that matter, where atmosphere is thin or virtually nonexistent. On Earth, however, thicker atmospheres pose a variety of problems.

"You're trying to light a match in a hurricane to get an engine to start," Frank explained. But with rapid design iteration now embedded in the culture, high failure rates were accepted so long as they led to quick improvements.

"Building many rockets allows for successive approximation," Musk later posted in 2020. By effectively lowering the stakes of failure, and by focusing on lower-risk satellite payloads, where the launches could serve dual purposes, the company could move faster, while opportunities to make money, or build infrastructure, were not to be squandered.

"I think what was really smart about the program overall was the primary mission was always to launch the satellite," explained Frank. "Once you had a satellite launched in a first stage that's coming back anyway into the atmosphere, it's a great opportunity for low-risk experimentation. So where you can't tolerate risk on primary mission, there you can tolerate a lot of risk."

In those early days at Vandenberg, however, the military needed a bit more convincing.

The idea of reusability had not quite taken hold.

"Please be comfortable with a large rocket flying at your base," Frank explained of his conversations with base officials. Somehow, however, they were ultimately convinced. And by 2015, SpaceX had signed a five-year lease for Cape Canaveral's Launch Complex 13, a launchpad at the air force station, adding to the Complex 4 West pad at Vandenberg. Then, on April 8, 2016, the company would make history. About eight minutes after launch from Cape Canaveral, while it was carrying an inflatable habitat into orbit, the first stage of a Falcon 9 spun around in mid-flight after separating and exercised a controlled power descent toward a drone ship floating in the Atlantic Ocean.

"It didn't put a hole in the ship or fall over, so we're pretty excited," Musk told reporters, later describing the feat as "trying to land on a postage stamp out there." SpaceX had proven it could land down-range on a ship that was bobbing up and down in the water, and it seemed as if the industry itself had turned a corner, as the emphasis on lowering costs continued. The barge had been rented off a website akin to "rentalbarges.com," Frank added.

"That's not actually the name of it, but it was the equivalent," he clarified. "We also bought lots of pipe from pre-insulatedpipe.com [and] usedairplanetires.com." Rather than buy new, SpaceX was looking for deals. In one instance, the company purchased fuel tanks on eBay from a chemical company, which had to be certified for structural integrity, reliability, and safety. It didn't always work. But in shepherding the design effort to get used parts through certification, the team was making progress, even despite some very public failures.

"We blew up a lot of stuff in 2014," Frank explained, again laughing. In fact, SpaceX had failed five times before that April landing at sea. A series of procedures typically follow a rocket landing, including safety assessments and recovery operations. Engineers pore over engine data to better understand precisely how the vehicle performed. But at least initially, when the rocket finally landed on that barge out at sea, the data would have to wait. Much of the rest of the company was basking in the moment. Celebrations had erupted.

"Texas proceeded to be really, really drunk," Frank explained. "They answered the phone and said, 'Whoa, congratulations, recovery. All right, bye,' without giving us any meaningful data."

"Hey, avionics," Frank recalled of a phone call he had made following the landing, telling the corresponding team, "I don't have permissions to do this step, because we've never done this step before."

The reply: "You guys are hot. Bye!"

"So that was actually quite a funny evening of just trying to do work with a company that was very drunk." Frank would later move to Hawthorne, for a job as SpaceX principal engineer and launch projects manager, where he worked on the nearly four-hundred-foot-tall Starship designed to eventually take humans to Mars. "About half my time at that point was doing the napkin sketches of what a Mars rocket launchpad would look like, how it should be, and just trying to get the general physics to work," he said. The two-stage rocket comprised a booster and an upper stage meant to carry crew and cargo over the months-long journey to the Red Planet, using a concept that required multiple launches. It was a model that Chinese leadership was clearly emulating. And yet while that happened, Frank and Lee had started thinking beyond rocket systems entirely.

The new focus aligned more closely with those of people like Kam, who wanted to spend more time thinking about space operations and infrastructure, and not just what it took to get there. Low Earth orbit was rapidly becoming the next factory floor. Space was bifurcating into both a frontier and a foundation for entirely new systems that were now coming online.

Together with a man named Vojtech Holub, a PhD in computer and electrical engineering, Lee would go on to found a Colorado-based company called ThinkOrbital, employing advances in electron beam welding to begin to solve one of the many problems found in manufacturing in the near vacuum of space, where traditional notions of conduction and convection do not apply. These were the kinds of developments that could usher in the kinds of megaprojects in space that would make George Lucas proud, using space-age welding guns to produce a stream of high-energy, high-speed electrons that could fuse metals or cut them apart. Frank would later join them as the company's chief technology officer.

Yet their mandate was more than just building, as Lee later explained. ThinkOrbital was also developing a space-based X-ray technology that had drawn the interest of the U.S. national security sector, effectively enabling operators to peer inside satellites from afar.

"Can you know [what actually] is inside Russian satellites?" I asked.

Lee smiled.

"That's a great question," he replied.

"One of the other effects is that if you take that stream of electrons . . . hit a dense piece of metal and slow it down, you generate a lot of heat, but you also generate X-rays. At first, we're like, great.

We've got our own X-ray source. We can inspect the welds that we just did."

But from his old air force days, he also wondered if there were other security applications.

"I was up at the National Space Council briefing, and they said, 'Well, you know, we have this problem. We believe that the Russians are hiding capability inside of satellites. How far away could you be [to employ X-rays] so it's not a provocative event . . . to do inspection and characterization of satellites?' "

"So we went back to the drawing board. Now we can use different types of X-ray sources. . . . I'm not going to tell you the how, because that's pretty company proprietary, but the itinerant effect is we can focus those X-rays down to get it onto a target and basically look inside a satellite, penetrating up to several centimeters of the skin on the satellite, looking actually through the satellite [from] about ten kilometers [away], to tell if there's something in that satellite that shouldn't be there," including the alleged presence of nuclear weapons deployed in orbit.

Once again, civilian technologies in space had proven useful in preparations for war.

· CHAPTER 28 ·

# THE ART OF WAR

· · ·

All armies prefer high ground to low,
and sunny places to dark. . . .
You should occupy the raised and sunny spots,
and there wait for him to come up.

—SUN TZU, *THE ART OF WAR*

More than two millennia ago, a Chinese strategist and philosopher wrote a military treatise that would outlast empires to be taught in modern college classes. Sun Tzu's *Art of War* was not just a manual of tactics but a timeless guide of strategy and philosophy as it relates to conflict, leadership, and decision-making. However, there is a lesser-known strategist who might be equally relevant here. A philosopher and tactician in his own right, Zhuge Liang was born in the twilight of the Han dynasty and rose to prominence during China's Three Kingdoms period, from AD 220 to AD 280. An adviser to Liu Bei, then a prominent Chinese warlord in the Eastern Han dynasty, Zhuge came of age at a time when rival kingdoms evolved and sought to monopolize political influence. Over the decades, warlords and statesmen from these three competing kingdoms would shift alliances and launch battles against one another in a power vacuum left by the crumbling of China's second imperial dynasty. It was a time of political upheaval, fragmented rule, intrigue, and conflict, and it was also fertile ground for many a novelist.

As such, there is a mix of history and folklore here. And it is not always easy to distinguish between the two. Zhuge, for instance, had a certain ability to "read the stars," according to legend. But there were other tales that seemed more grounded. In one such account, Zhuge found himself surrounded by enemy forces just outside the garrison walls. What happened next is a bit murky, but according to legend Zhuge—also known by his courtesy name, Kongming—perhaps knowing he was outmatched, launched an early form of aerial signaling, later known as the Kongming lantern. These paper balloons have a relatively simple design: an open bottom where a candle is lit. Heat from the candle generates lift by causing the air molecules to move faster and spread out, compared with the denser, cooler air outside, which as a result makes the lantern rise.

Kongming reportedly instructed his men to assemble these balloons by the thousands inside the confines of the fortress and release them into the air. The night's sky soon became awash with floating lights, perhaps a signal for help or an attempt to create the illusion of a more technologically advanced force waiting inside those walls. Either way, the anticipated attack never came. The garrison had been spared. And sky lanterns had become some of the first wartime aerial communications, a usefulness not lost on future generals, both in its signaling and in its deception.

While Sun Tzu wrote about the imperatives of both, it wouldn't be until centuries later in a much different part of the world that the renowned Prussian military theorist Carl von Clausewitz would be more explicit in his 1832 book, *Vom Kriege,* or *On War*. In that seminal work, Clausewitz determined that "three quarters of the factors on which action in war is based are wrapped in a fog of uncertainty." Incomplete information, misinformation, and confusion are therefore not temporary dynamics of conflict but enduring features to be recognized and managed as persistent elements.

Fast-forward to modern warfare, and it is now satellite constellations and their supporting cyber infrastructure that are effectively the new sky lanterns, capable of piercing the "fog of war" or generating it, while providing real-time intelligence on enemy movements, supply lines, and weather conditions, and also detecting through AI-run sensors a range of infrared and other signals invisible to the human eye. This is a brave new world in battlefield awareness, woven as it is "into the fabric of modern military operations," as the

U.S. lieutenant general Steven L. Basham later put it. And investments began to point to that understanding. By 2023, the U.S. Space Force had nearly doubled its budget. A year later, global government space spending alone would reach a record $135 billion, propelled mostly by defense-related budgets, with the United States accounting for more than half of that spending (59 percent). And yet by the summer of 2025, President Trump's signature domestic policy legislation would go further, initially devoting at least $150 billion to build (along with military shipbuilding efforts) a new kind of space-based missile defense system, now commonly referred to as Golden Dome, something Defense Secretary Pete Hegseth later called "a generational investment in the security of America and Americans," even as questions emerged about how those funds would actually be dispersed. The scope of the plan came with broad strokes, and was described as a layered-defense shield, replete with a network of space-based sensors and interceptors, oriented against air- and spaceborne threats, including non-kinetic means, such as electronic warfare and cyber capabilities, as well as more advanced threats, such as hypersonics, which are defined as objects traveling faster than Mach 5, or five times the speed of sound.

Again, it was clear space was an active battleground, supported by ever more precise and autonomous machines in orbit. And yet it wasn't just about imagery anymore, at least not in the traditional sense. Surveillance had evolved well past mere snapshots of Earth. When I spoke with Dan Smoot, chief executive officer at Maxar Intelligence, now known as Vantor—which contracts with governments and other entities to deliver AI-driven, real-time monitoring—he emphasized that "it's about how you coalesce that data to bring out differentiations.

"What people are really interested in is what's changing. It's not just about that one image. It's about the ability to have multiple revisits into an area," he said, "to bring more intelligence to everything from a defense and intel perspective, but also from a civilian perspective.

"Being able to see that [proverbial human] ant farm form in a different way is critical," especially in the context of growing efforts to forge a kind of "digital twin of the world," or a virtual representation used to simulate a barrage of scenarios to assess real-world situations and their prospective outcomes. Everything from global trade and climatic patterns to holiday shopping habits and enemy troop positions was on offer. A 2024 McKinsey & Company report found that the global market for digital-twin technology was expected to swell by roughly "60 percent annually," reaching $73.5 billion by 2027. The goal was effectively to build increasingly accurate AI-powered simulations, born of high-resolution satellite imagery and geospatial data that could replicate cities, forests, oceans, and other large-scale systems, with increasing amounts of that data processed on orbit. But the notion also raised major questions of privacy. A maturing digital-twin technology could afford corporate or geopolitical rivals, like China, critical insights into Western infrastructure, resources, and other sensitive assets, or identify exploitable vulnerabilities in supply chains, while giving authorities and companies new powers to monitor their citizens and consumers alike, effectively painting a revolutionary picture of the human species that reveals in three-dimensional ways how we interact with unprecedented precision and speed.

"In some cases, [operators] cared about [receiving data and insights] in a ninety-minute time frame," Smoot explained of his customer base. Operational tempos were increasing, though not just in the acquisition of high-resolution images but also in the processing and extracting of meaningful insights from such enormous amounts of new data. The elephant in the room, however, was still that distinctly human question: How long could we mere people stay in the loop? It was a question I put directly to Smoot.

"It's one thing to provide certain information and get it into people's hands," he replied. "It's another thing to get that information and deal with that last mile of decision-making. There has to be a human element when it comes to certain types of decision-making when you're talking about defense and intel." But, he said, "we are a little bit challenged when our technology moves faster than our processes and policies."

With the dawn of new threats like hypersonic weapons radically compressing the time afforded to decision-makers, it wasn't clear where humans would ultimately fit in. Growing speed requirements seemed like too much to overlook. To put those national security risks in context, Deborah Lee James, Ryan McCarthy, and Michael E. White of the Atlantic Council Hypersonic Capabilities Task Force explained how "to deliver effects on a target at 500 miles, a traditional subsonic cruise missile, such as the Joint Air-to-Surface Standoff Missile or Tomahawk, would take approximately one hour of flight time. Hypersonic missiles can make that trip in less than 10 minutes," with the added ability to undertake evasive maneuvers, change course, and fly at low altitudes—as opposed to the more standard parabolic arc of conventional intercontinental ballistic missiles. Current missile defense systems, such as the Patriot missile system or THAAD (Terminal High Altitude Area Defense), suddenly seemed woefully inadequate, especially in an era where orbital data centers, leveraging AI-powered sensor fusion, could process and address emerging threats in real time, often at the very edge of space where those threats are commonly first detected.

Many of the core aspects of these technologies, however, weren't in fact all that new. By the 1950s, the U.S. Air Force and NASA had begun hypersonic testing with the X-15 program, an experimental aircraft that would reach speeds of Mach 6.7, and later cruise missiles as well as scramjet engines capable of tolerating the extreme temperatures of hypersonic flight. But with increased focus on counterterrorism efforts, and the lack of a near-peer competitor in the wake of the Cold War, the United States—once considered a pioneer in hypersonic technology—had turned its focus elsewhere, only to be outpaced. Both Russia and China are thought to have leapfrogged technologies, particularly in the development of hypersonic glide vehicles and air-launched hypersonic cruise missiles. Russia's Oreshnik missile—Russian for "hazelnut tree"—travels at ten times the speed of sound and has already been employed on the battlefields of Ukraine. In fact, Russia, China, North Korea, and Iran are all developing hypersonic variants. Automated responses using advanced tracking, space-based sensors, and layered-defense systems suddenly seemed among the few ways to close those yawning security gaps.

And yet as contractors looked to meet these new "Golden Dome," multilayer security efforts, the broader paradigm was again quietly shifting. Beyond tracking hypersonic weaponry, interest was growing in picking apart the invisible scaffolding that underpins much of our shared reality, including those in contested locales. Here, digital twins were effective tools, but they also were only at their inception. Frontier concepts such as space-based quantum sensing promised to unveil a range of an even more profound sense of surveillance, with a radically more comprehensive picture. This included tracking and sensing objects based not on emitted signals or visual objects but through the natural forces they exert on the surrounding environment.

That is to say, one cannot hide from gravity.

Anything with mass, however small, exerts a corresponding gravitational force. And with emerging quantum tools, detecting minute shifts in gravity fields can reveal hidden truths about objects, where they're located, and what anomalies they create. Picture, for instance, militaries with the kind of centimeter-level precision that could dramatically enhance navigation and surveillance, evaluating enemy launch sites with new levels of detail. This kind of precision is hard to fathom from a human cognition perspective, but consider instead AI-powered systems driving analysis and ultimately policy recommendations with the kind of relentless, real-time clarity that the processing of incomprehensibly vast data sets might yield, albeit with questionable long-term levels of accountability, ethics, and alignment with human intent.

Such applications also promised to extend well beyond the battlefield. Certainly, the potential effects in space were profound. At a more immediate level, scientists could use such measurements to scour asteroids for rare metals, or hidden caverns on the Moon and Mars suitable for human settlement. Focused back on Earth, such quantum instruments could also provide farmers with inch-by-inch insights into soil composition and moisture levels, or miners with an ability to locate valuable minerals without ever visiting the site.

"It's an exciting prospect," JT Janssen, chief scientist of the National Physical Laboratory in the U.K., told me. "We'll measure more, and learn more," including such cosmic enigmas as dark energy.

It was groundbreaking stuff. Now the question was how those technologies could be applied from space in more practical use cases. Would stealth fleets lose their cover? And how might advanced nations use the technology more broadly to exploit countries less informed about their own resources? The tools of both commerce and war were rapidly changing. Yet beneath this revolutionizing transformation in both civil and military surveillance lay a far more fundamental and enduring legacy of space weaponry.

It was a phenomenon that every nation would eventually have to confront.

· CHAPTER 29 ·

# TURNING POINT

By 2007, the space world was on the cusp of a new era. A barely two-year-old Joint Space Operations Center at Vandenberg Air Force Base had been handed a monumental task: detect, track, and catalog every man-made object in orbit—a growing swarm of everything from satellites to spent rocket stages. B. Chance Saltzman, then only a lieutenant colonel—though the future chief of space operations for the yet-to-be-formed U.S. Space Force—was at the time working in a small office on the third floor.

"Our job was to collect statuses, mostly the space domain awareness—we called it space situational awareness back then," Saltzman reportedly explained. But even back then, one could sense a geopolitical shift was underway. He knew at the time, for instance, that Beijing had rolled out a multistage, solid-fuel, medium-range ballistic missile to its launchpad at Xichang Satellite Launch Center. What he didn't know was that the rocket's payload also included a kinetic kill vehicle to be directed at one of China's aging polar orbit weather satellites. This was not just a tech evaluation. It was a demonstration of China's newfound power.

"To be honest, I think there was this feeling that it was just going to be a test."

Shortly after launch, however, the vehicle sliced through the atmosphere and slammed into the satellite, traveling at twenty thousand miles per hour. At that pace, it didn't need explosives. It simply ripped through the Fengyun satellite, resulting in a debris cloud that quickly formed in orbit.

"I remember very clearly, the radar operator who's looking at his chat rooms connected to the ground radars spread around the world, and he looked back over his shoulder and he said, 'We have multiple headcount.'"

"Headcount" is a reference to a fragmentation event, whereby a single object in space breaks apart into multiple pieces. The explosion quickly transformed into a ring of debris that spread out across much of low Earth orbit. Thousands of pieces of space junk were now moving at several orders of magnitude faster than a bullet, imperiling other satellites and spacecraft: a swirling metallic cloud that was circling and colliding and changing trajectory.

"In that moment, I knew that things had irrevocably changed—space was no longer safe," he said.

In fact, there were more than three thousand pieces of debris, at least 97 percent of which were "significantly sized," as described by the later-formed U.S. Space Surveillance Network (SSN). And they imperiled anything in their way. SSN can track objects as small as ten centimeters in diameter, a bit larger than a baseball, and can provide information for spacecraft to take avoidance maneuvers, if necessary. The problem was, however, that explosion also created tens of thousands of golf-ball-sized debris (and smaller) that were now whipping around Earth, where many satellites—as well as the ISS—were operating. And at those speeds, even a fleck of paint could pierce a space suit. And much of it would remain there for decades, a persistent and growing threat to near-Earth orbiting satellites and what had become a steady increase of launch vehicles. The United States had also contributed to space debris with the 2008 destruction of a failed U.S. satellite by an Aegis Cruiser, albeit at considerably lower altitudes, whereby most of the debris was believed to have burned up upon reentry. The Chinese test, by contrast, was not only in higher orbit, which meant the fragments would remain for far longer, but also positioned as a show of military capacity: a wake-up call that prompted the United Nations to issue a series of debris guidelines, which included end-of-life satellite disposal, while also marking a decided shift in military thinking.

In coming years, concerns over the threat of anti-satellite capabilities seemed only to intensify. And by May 2024, the Pentagon spokesperson Major General Pat Ryder asserted that Moscow had launched "a satellite into low Earth orbit that we assess is likely a counterspace weapon, presumably capable of attacking other satellites in low Earth orbit." Roscosmos offered few details about the launch, only that a Soyuz rocket carrying a spacecraft had blasted off from the Plesetsk Cosmodrome "in the interests of the Ministry of Defense of the Russian Federation." A trio of unnamed sources told NBC News that Russia was in fact developing a nuclear space-based weapon capable of targeting American satellites, which encouraged Mike Turner (R-Ohio), then-chair of the House Intelligence Committee, to press to declassify related information. If true, the act would violate the 1967 Outer Space Treaty, which expressly prohibits nations from placing "nuclear weapons or other weapons of mass destruction in orbit, or on celestial bodies, or station[ing] them in outer space in any other manner."

"I don't think we should respond to any information hoax from Washington," Deputy Foreign Minister Sergei Ryabkov told Russia's news agency Tass in response to the report, calling it "fake news."

"The Americans can say whatever they want but our policy does not change from this," said Ryabkov, noting that the Kremlin has "always consistently opposed the deployment of strike weapons in low-Earth orbit." If such an attack were carried out in orbit, it is hard to overstate the consequences, considering how deeply dependent modern society has become on its space systems. In some ways, however, humanity wouldn't have to guess. It had happened before, albeit during a much different time.

To understand precisely when, how, and why, one must wind the clock back to an evening in the summer of 1962.

· CHAPTER 30 ·

# STARFISH PRIME

In a lonely stretch of the Pacific Ocean, devoid of a permanent population and home only to a rotating crop of American military and government scientists, the Sun had long dipped below the horizon. The clock had recently struck 11:00 p.m. Honolulu time on July 9, 1962. But for the moment, it might not have seemed that way.

"It looked like noon," a man named Greg Spriggs told *National Geographic,* describing the view from his perch on Midway Atoll where he watched with his father, nearly a thousand miles away from where a new, and devastatingly powerful bomb had been detonated some 250 miles above Earth. Witnesses as far away as New Zealand reported "rainbow stripes" as nuclear auroras illuminated an otherwise pitch-black night. The United States had just conducted the largest nuclear test in space, dubbed Starfish Prime, which unleashed the power of a 1.4-megaton weapon.

"It looked as though the heavens had belched forth a new sun that flared briefly, but long enough to set the sky on fire," as described by one account published in the Hawaii-based *Hilo Tribune-Herald.* The biggest of several high-altitude nuclear tests conducted by both the United States and the U.S.S.R. during the 1950s and 1960s, Starfish Prime was several orders of magnitude larger than the bomb dropped on Hiroshima, delivering an electromagnetic pulse that prompted blackouts across the region. Carried atop a Thor intermediate-range ballistic missile, the weapon, through its blast, also destroyed roughly one-third of the twenty-four satellites in orbit, with accompanying radiation degrading the remaining satellites in the years to come, raising the intensity of Earth's Van Allen belt "by several orders of magnitude."

The weapon had proven far more devastating than expected. Perhaps in recognition of that terrifying new power, U.S. President John F. Kennedy and Soviet Premier Nikita Khrushchev would soon after sign the Limited Test Ban Treaty, effectively prohibiting atmospheric and exoatmospheric nuclear testing, before the two nations eventually also signed the Outer Space Treaty, which effectively banned nuclear weapons in space. A new atomic awareness was growing, even if the space systems back then were not nearly as vulnerable as their modern counterparts.

Today, the global economy relies on thousands of satellites, a reliance that has spurred new debates about norms of behavior and the urgent need for a more comprehensive global acknowledgment of such modern realities, especially amid reports of a potential Russian nuclear weapon in space. The prospects of massive disruptions to everyday life now loomed large, threatening global communications, navigation, and even timing services.

In fact, virtually all of U.S. critical infrastructure, and much of the world's infrastructure, depend on the U.S.-run Global Positioning System (GPS), a service that has over the years woven itself into most operations because of its high degree of reliability and availability, while remaining relatively secure because of the precious few alternatives.

That, of course, was then.

With the rise of Chinese, Russian, and European competitor systems, that old aura of protection has now worn thin, whereby an attack on GPS no longer equates to a simultaneous hobbling of an attacker's own systems and infrastructure. A widespread GPS outage could thus trigger cascading effects as it pertains to supply chains, emergency services, financial services, agriculture, and even energy and water management systems.

And yet in space, it doesn't take a nuclear blast, or even a kinetic strike, to wreak havoc on satellites and spaceships. In fact, it might not take an attack at all. NASA lists orbital debris—along with micrometeoroids—as the "number one risk for NASA's human spaceflight programs." And the effects pose terrestrial dangers as well. In March 2024, for instance, a two-pound hunk of space junk from the ISS entered our planet's atmosphere and smashed through a family's home in Naples, Florida. The debris was a leftover pallet loaded with spent batteries, which had not quite burned up in reentry. "The NASA Kennedy team initially evaluated the photos from the homeowner and noticed key features still present on the debris similar to a stanchion from the NASA flight support equipment used to mount the batteries on the cargo pallet," explained Joshua Finch, a Kennedy Space Center spokesperson, in a reported statement. No one was injured. But the family sued, highlighting at least two key developments: (1) the rise of space law as an emerging legal field, particularly amid commercial expansion and emergent questions about the obsolescence of existing legal frameworks, and (2) the growing challenge of space debris.

Years later, in November 2025, three Chinese astronauts were traveling aboard their Shenzhou-20 spacecraft when it was struck by a piece of space debris traveling at roughly eight kilometers per second, cracking one of the craft's windows. It was clear orbital junk was becoming an ever more serious hazard, and one that was only expected to worsen given the sheer volume of machinery now being launched. It was also something Europe, in particular, was now racing to address before it became too late.

· CHAPTER 31 ·

# SPACE JUNK

Nestled up against the Dutch North Sea coast, the seaside town of Noordwijk is known both for its rolling sand dunes and as an incubator of European ambitions in the cosmos. Small and unassuming, it is in fact the technical heart of the European Space Agency (ESA), a place that draws space minds from across the Continent. Among them was a Portuguese-born engineer named Tiago Soares, who—by the time we spoke—was working as the agency's Clean Space co-initiator and lead on something called Zero Debris. In essence, it called for an end to the unmitigated accumulation of junk in space by 2030, with the idea of bringing new standards to that which dangerously circled the planet. In many ways, despite the sheer expanse of Earth's orbits, the situation was growing more troubling.

"We are already beyond the threshold of sustainability," he told me one evening. "Our simulations show we are already above the threshold at which even if we stop launching completely, the [amount] of debris [that's orbiting] will increase by itself."

That bears repeating. In essence, even if humanity never launched another rocket or sent another satellite into space, the debris field around Earth would continue to grow, Tiago explained. Now the issue was more about mitigating those effects. For context, debris circling below orbits of 600 kilometers generally return to Earth and, depending on their size, burn up in reentry. Above 800 kilometers, that orbital decay can take centuries, while more than 1,000 kilometers can result in an orbit of more than a thousand years. Derelict satellites, spent rocket stages, fragmentations during launches, structural failures, micrometeoroid strikes, and anti-satellite strikes were all to blame. In low Earth orbit, an area typically defined as between 160 and 2,000 kilometers, no fewer than 26,000 softball-sized, or larger, pieces of junk circle the planet every ninety minutes, give or take depending on the orbit. And there are more than 500,000 pieces roughly the size of a marble that make that orbital journey. More than 10 million pieces the size of a grain of salt are also in orbit. The latter, though small, given their speeds of roughly 17,500 miles per hour, could still puncture a space suit or damage a satellite. Even without additional launches, the chain reactions of continuously colliding objects in orbit bring the possibility of creating even more circling debris.

That scenario was in fact proposed back in 1978 by the NASA scientists Donald J. Kessler and Burton G. Cour-Palais, and has since become known as the Kessler effect or Kessler syndrome—a concept that points to ever larger shrapnel fields zipping around the planet much to the worry of astronauts and satellite operators alike, so much so that eventually it becomes perilous to the point of making space inaccessible.

"Satellite collisions would produce orbiting fragments, each of which would increase the probability of further collisions, leading to the growth of a belt of debris around the earth," the two men wrote in a study titled "Collision Frequency of Artificial Satellites: The Creation of a Debris Belt."

Nearly half a century later, that so-called belt included 12,400 tons of man-made objects. With nearly three thousand defunct satellites and other craft careening around Earth, Tiago and others at ESA were now effectively saying that the Kessler cascade had already begun. It would take decades for this runaway effect to significantly impede human abilities to conduct space activities, however, given the size of the areas and relative densities of existing debris fields. And yet that process had seemingly started.

But it wasn't until 2009, a full two years after China's anti-satellite test, that global leadership seemed to truly begin to take the Kessler effect more seriously. That year, in February, nearly eight hundred kilometers above the Taymyr Peninsula, an Iridium communications satellite and a derelict, nearly two-thousand-pound Russian Strela-2M class orbiter swooped over Siberia on a collision course in excess of twenty thousand miles per hour. Launched in 1993, the retired Russian satellite, known as Cosmos 2251, had an operating life of only about five years and was long considered inoperable. It could not, in other words, be maneuvered out of the way. Yet despite tracking from both U.S. and Russian armed forces, there was reportedly little if any warning of the coming collision. When the two smashed into each other, some two thousand shards of debris, measuring at least four inches in diameter, and countless more less than that size, were strewn around Earth. It was one of the worst collisions in space history.

Tiago, meanwhile, was at work.

He had just begun working as a system engineer at ESA and would watch as the messy aftermath unfolded. As the fragmentation cloud expanded and multiplied in what NASA described as "a very large, very energetic event," the finger-pointing would soon begin. The Russians noted that Cosmos 2251 was a derelict satellite, blaming the event on the Virginia company's failure to change course, while Iridium essentially maintained that it was not required to maneuver out of the way. Questions of liability would soon follow.

"I remember at that time everybody started talking about why it happened, of course," Tiago recalled. "Whose fault was it? Why didn't the active satellite move? And the responsibility of what happens after. There was a lot of debate."

Low Earth orbit had suddenly been made more dangerous as metal shards whipped around Earth, compelling evasive maneuvers of orbiting spacecraft. The effects were profound, particularly for U.S. Strategic Command and the U.S. Joint Space Operations Center, both of which began tracking constellations and orbiting space objects more closely with advanced warnings initiated within seventy-two hours for close approaches. Meanwhile, the field of space situational awareness—or the capacity to track and predict the location of both man-made and naturally occurring objects in orbit—began to take on greater urgency.

Tiago was focused on the bits that didn't stay in orbit, and yet also didn't fully burn up on their descent through Earth's atmosphere—shrapnel from the collision that was now headed to Earth.

"I started using tools to predict the risk on the ground," he explained, as well as growing threats to satellite operators' mobile and GPS networks, which later became part of the ESA's Clean Space Initiative. "That's where it started."

But, noted Heiner Klinkrad, head of ESA's Space Debris Office, "within a few decades, there are going to be collisions among large objects that will create fragments that can do further damage. The only way to keep this from happening is to go up there and remove them. The longer you wait, the more difficult and far more expensive it is going to be."

In 2023, it nearly happened again when the space object tracking firm LeoLabs noted that a Soviet atmospheric research satellite launched in 1976 missed slamming into a forty-four-hundred-pound stage of a Long March 4C rocket by just 188 feet. A year later, NASA's Thermosphere, Ionosphere, Mesosphere, Energetics, and Dynamics mission spacecraft, whose mission included examining the Sun and Earth's upper atmosphere, came even closer—an estimated 66 feet from smashing into another derelict Russian satellite, Cosmos 2221. LeoLabs called it "too close for comfort." And as recently as October 2024, a Boeing-made communications satellite exploded from a so-called anomaly in orbit, which compelled the tracking of at least twenty pieces of newly formed space junk, along with countless other bits of micro-debris.

The problem was clear, and had all the makings of getting worse. If it did, that steady accumulation of orbital debris could eventually jeopardize the very infrastructure this new space era sought to build, while threatening future launches. In the United States, however, orbital oversight seemed only to be worsening. As part of the sweeping government-wide cuts first triggered by the Department of Government Efficiency, initially under Elon Musk, the lesser-known Office of Space Commerce within the National Oceanic and Atmospheric Administration saw staffing cuts of as many as sixty civil servants and contractors. This included reductions to the team working on a Traffic Coordination System for Space to handle basic space situational awareness data, as well as services to civil and private space operators in support of spaceflight safety. That team had been tasked with cataloging orbit objects and issuing warnings for close approaches among satellites. And they were losing capabilities at the precise moment orbital traffic was increasing. More broadly, senior NASA staff wondered openly about longer-term effects on the agency amid such workforce cuts, staff buyouts, and early retirements.

"As a 35-year NASA veteran, I can reasonably say we as an agency haven't had the challenge we faced with today, at least over my career," Steven Hirshorn, NASA's chief engineer for aeronautics, posted on LinkedIn on July 24, 2025. "As a result of attrition from the Deferred Resignation Program, Voluntary Early Retirement Authority, and other departures, our ranks of engineers, managers and leaders will shrink by ~25%, according to the latest numbers. And yet, NASA's mission continues, and expectations for success (including our own) remains.

"It's quite the challenge," he added. "How do we move our missions forward with such experience loss and do so with the excellence NASA is famous for?"

Yet staffing cuts and tracking debris were only part of the new paradigm. Equally pressing was the unresolved question of how to actually remove debris from orbit.

Space apparently needed its garbage men (and women). And in this commercial era, venture-backed proposals had already started to flood in. In fact, by February 2024, a spacecraft built by a Japanese company called Astroscale became the first mission designed to drag a derelict rocket out of orbit. It was an attempt to make good on UN regulatory mandates to remove space objects a quarter century after their expected lifespans, which in practice included a lot of hulking Soviet craft like the one that collided with Iridium 33. But it wasn't so simple. Not only were there questions of how to make a deorbiting business profitable, but these objects were also uncontrolled and devoid of communications. There was also the question of ownership, and whether derelict satellites produced by one country could be hauled down by companies from another.

Technically, the 1972 Convention on International Liability for Damage Caused by Space Objects had determined that whoever put the satellite up there is ultimately responsible for getting it down. And in limited ways, there was some enforcement. In fact, in 2023, the U.S. Federal Communications Commission (FCC) handed out the first-ever space junk fine to DISH Network, regarding a twenty-one-year-old communications satellite called EchoStar 7, after its end-of-mission disposal orbit was "well below the elevation required by the terms of its license," the FCC said. The fine was $150,000.

But that kind of thing was the exception.

In general, not only was accountability lax, but geopolitics would also factor in what technologies would be used. Photon pressure, ablation techniques, and lasers to rid the orbit of space trash could also be employed as space weaponry, which was politically fraught and therefore impeded its adoption—especially given the long Cold War history up there. Debris collection efforts would have to be mindful of the past, lest they prompt a new arms race in space.

Yet signs suggested it may have already begun.

The threat, however, was not just in orbit. A growing global launch sector, which had created this new, crowded, and contested reality in space, also had implications for environmental and human health impacts back on Earth. For instance, when Starship's second test flight ended in an explosion eight minutes into the mission, with a self-destruct system on its upper stage triggered shortly thereafter at an altitude of about 150 kilometers, the effects blew open what some researchers called one of the largest "holes" ever seen in Earth's atmosphere. Releases of aluminum oxide nanoparticles into the upper atmosphere could "endure for decades" and lead to "significant ozone depletion," according to a NASA-funded 2024 study published in *Geophysical Research Letters*. "Due to their small size, the byproducts of spacecraft reentry can endure in the atmosphere and remain unnoticed until ozone concentration levels start decreasing." The ozone layer effectively filters out most of the Sun's harmful ultraviolet radiation. So its depletion can increase the risk of skin cancers, eye cataracts, and immune deficiency disorders, while also affecting both terrestrial and aquatic ecosystems as it relates to food chains and biochemical cycles, effectively stunting plant growth in ways that could reduce agriculture.

The red flags were clear. As we reach for the stars, we had better be careful to not erode the very veil that protects life on Earth.

Of notable concern in that aforementioned study, and in what had constituted the first atomic-scale simulation of aluminum ablation during reentry, researchers found the "demise of a typical 250-kg satellite can generate around 30 kg of aluminum oxide nanoparticles," which could catalyze reactions that destroy ozone. It determined that aluminum oxide compounds generated by the entire population of satellites reentering the atmosphere in 2022 were estimated at around 17 metric tons, which had caused "a 29.5% increase of aluminum in the atmosphere above the natural level," finding that "mega-constellations point to over 360 metric tons of aluminum oxide compounds per year."

Satellite constellations, meanwhile, were expected to grow by orders of magnitude, backed by a launch sector that was also seeing exponential year-over-year growth.

And yet it wasn't just aluminum oxide. Nitrogen oxide, chlorine, hydrochloric acid, black carbon, and soot were also generated with launches, with National Oceanic and Atmospheric Administration scientists determining that upper atmospheric conditions were at times "peppered with particles containing a variety of metals from satellites and spent rocket boosters vaporized by the intense heat of re-entry." The findings, published in an October 16, 2023, *Proceedings of the National Academy of Sciences* report, found that such particles matched the "ratio of rare elements" used in the special alloys for rockets and satellites. What's more, those same pollutants released in the upper stratosphere tended to linger longer than those in the lower atmosphere, which meant a single rocket's emissions could have a greater and more lasting effect than that of even several commercial airliners, particularly when those rockets employed kerosene as a propellant.

The launch industry, in other words, now seemed to be at risk of endangering one of Earth's few climate success stories. Whereas questions about the ozone layer had long been a concern, since 1987—after nearly two hundred countries pledged to do away with ozone-depleting chemicals, namely chlorofluorocarbons—the planet had seen a gradual recovery.

To be fair, efforts in the space industry were under way to explore more environmentally friendly and health-conscious propellants, including oxidizers like ammonium dinitramide, which when heated decomposes into nitrogen, oxygen, and water. That was in contrast to highly toxic compounds like hydrazine, long suspected of contributing to abnormally high rates of hormonal and blood disorders around the Baikonur launch complex in Kazakhstan.

And yet beyond atmospheric impacts, there were also questions of how such an increased cadence of launches might affect other aspects of life. In August 2024, researchers with *Consumer Reports* released the results of 196 food samples, and discovered 67 percent of them contained perchlorate, a chemical often used in rocket fuel, as well as in fireworks, certain plastics, and airbags. The report found "in general, baby/kid food, fast food, and fresh fruits and vegetables had the highest levels, with children's foods averaging the highest average level, 19.4 ppb."

"For a child between 1 and 2 years old, a serving of the boxed mac and cheese we tested would hit nearly 50 percent of the EFSA [European Food Safety Authority] limit, and servings of the baby rice cereal, baby multigrain cereal, and organic yogurt we tested would each hit about a quarter of that limit," the report noted. "That means with one serving of each of the above foods—very possible over the course of a single day—a child would exceed the EFSA's safe daily limit." Excess perchlorate exposure has been linked to an increased risk of disrupting normal thyroid function, particularly thyroid hormone regulation of embryonic development, making fetuses, infants, young children, and pregnant women particularly susceptible to its effects. That said, the linkage to aerospace was highly speculative, and seemed questionably tied to an increased launch cadence. While perchlorate composite propellants were commonly used in solid rocket fuels, they were not typically used in liquid rocket engines, which is where the overwhelming amount of growth appeared to be happening. Many of those fuels, however, were nonetheless still considered greenhouse gases. And with no set global emissions standards, questions were now mounting as to what the long-term effects might be.

Yet the very nature of accountability was also changing. When, in early March 2025, for instance, a SpaceX Starship exploded during its eighth test, roughly two months after it lost the upper stage on its previous flight, the resulting debris field prompted a series of canceled commercial flights across Florida and the Caribbean out of concern for the potentially devastating effects of falling debris. There were no reported injuries, and yet the incidents also raised important questions more broadly for commercial space. Given SpaceX was a private company, investigations tended to work differently. After Starship explosions, the Federal Aviation Administration (FAA) effectively instructed SpaceX to investigate itself. FAA reviews those findings and then approves licensing for future flights when it is satisfied the issue has been sufficiently addressed.

That was in stark contrast to NASA incidents, in which detailed reports and instructions for corrective action were generally made available for broader public scrutiny. Therein, perhaps, lay an Achilles' heel of space commercialization: lack of a paper trail. During Starship's eighth test, the FAA's guidance was effectively the same. The agency, however, did respond with the following statement, which seemed to provide some qualifiers.

"A mishap investigation is designed to enhance public safety, determine the root cause of the event, and identify corrective actions to avoid it from happening again," it said. "The FAA will be involved in every step of the SpaceX-led mishap investigation process and must approve SpaceX's final report, including any corrective actions."

And yet for those champions of a more rapid pace of industrial growth, and those principally concerned with the speed of the race against China, not to mention the perils of public disclosures (that is, paper trails) regarding the IP of proprietary technology, adopting a NASA-style approach to regulations was not only ill suited but potentially dangerous. For them, too much emphasis on regulation would only hamper U.S. competitiveness at a time when American space efforts needed to be unleashed. To space executives with tech backgrounds, many of whom embodied that move-fast-and-break-things ethos often found in software development, the agency's old regulatory-laden approach was anathema to progress. Still, even in Europe, where regulations were often more developed, a growing sense of competition among companies—both with each other and on the international stage—was taking hold.

· PART II ·

# THE NEXT FRONTIER

· CHAPTER 32 ·

# ITALIAN ENTREPRENEURS IN THE RED DESERT

When I first met Luca Rossettini, he was pulling an astronaut's helmet over his head. It was the final piece of an otherwise complete simulator space suit at the Mars Desert Research Station, an analogue Martian habitat situated in a remote part of Utah, which—if you squinted—might have seemed just like the Red Planet. Days here began with the illusion of the world future Martian crews were one day meant to call home; along with an almost unspoken understanding that this wasn't pretend, but rather preparation for the place that would define the next chapter of human exploration; built on those early lessons that many hoped would start in earnest at the lunar south pole.

Helping Luca was a Moroccan-born researcher named Nadia Maarouf, a clinical scientist at the University of Calgary focusing on the impacts of prolonged microgravity on the human body, including bone density loss, muscular atrophy, plasma loss, and the upward shifts of fluid in the body as a part of research that examined the effects of space on cardiovascular health. It was a truly international crew. But it was Luca who stood out, not least of which because he was the founder and CEO of one of Europe's more prominent emerging space companies.

Having founded D-Orbit in 2011 after attempting to become an astronaut, he had earned his PhD in energetics from Politecnico di Milano, the largest technical university in Italy. It was there, after a brief stint at NASA Ames in California, where he said he began "recognizing the pivot to commercial space," having decided to create a company that devised solutions for that "last-mile" delivery of satellites and transportation in orbit, as well as the growing problem of space debris. The goal was to build orbiting stations that could disassemble aging satellites, recycle their parts, and conduct space manufacturing in orbit. And yet for the time being, tucked inside that cylindrical habitat, he just seemed to be soaking in the moment. With his helmet now secured onto his simulator space suit, the Italian founder was standing beside the facility's mock air lock, waiting to step into that red desert.

For those curious about life on the Red Planet, the Mars Desert Research Station provided a novelty glimpse. Situated just outside the remote town of Hanksville, where ancient riverbeds butt up against swirling layers of iron-oxidized rock, the region has a decidedly otherworldly feel to it. The facility itself is made up of six structures, painted white in contrast to the rust-colored countryside, which include six staterooms with bunks, as well as a seventh, which consisted of little more than a loft mattress a few feet below its conical roof. That, it turned out, was where I was to sleep. The options, both sleeping and eating, were limited. At one point, one of the crew handed me a candy bar made of dried crickets. Crunchy and protein packed, it seemed perfectly on brand for life on Mars. But it wasn't all spartan. Perhaps expectedly, those in the Italian contingent of the crew had also smuggled in a few bottles of olive oil, along with a few other culinary treats. Even on Mars, Italians (and those of us with Italian heritage) refuse to compromise on flavor.

That said, the real thing had some major supply impediments to overcome, given missions to Mars could take up to nine months. Nuclear-fueled propulsion and/or an Earth-to-Mars supply chain of Starships could begin to change that math, but for now those were more aspirational than reality based.

Mars was, to be frank, just a nasty place to live.

Median temperatures were equivalent to about as cold as it gets on Earth, with atmosphere about a hundred times thinner, and almost entirely comprised of carbon dioxide, with only traces of nitrogen, argon, and oxygen. $CO_2$ might encourage crop growth through photosynthesis, but real farming would still require some

Nadia Maarouf, clinical scientist at the University of Calgary, gears up at the Mars Desert Research Station, situated in a remote part of Utah that looked like the Red Planet.

Desert Research Station, among the largest and longest-running Mars surface research facilities, where two simulated Mars analogue habitats are operated by the Mars Society. Built near Hanksville, Utah, it has the look and feel of a locale that might resemble the barren Martian landscape. A team of analogue astronauts from Italy, France, and Canada conducted analogue off-Earth experiments related to Mars and operating in space more generally.

Analogue astronauts return to the "habitat" while operating at the Mars Desert Research Station in Utah, 2022.

Analogue astronaut and D-Orbit CEO/founder Luca Rossettini had a penchant for traveling farther afield than the rest of the team during an analogue mission, Utah.

basic degree of organic material. Crops would also need to be shielded, from both temperatures and radiation. In a peer-reviewed study published in the journal *Space Weather,* an international team of researchers found that while human travel to Mars may be feasible, staying longer than four years would in fact expose astronauts to fatal levels of radiation from solar energetic particles and galactic cosmic rays. Thick shielding on the ship's hull or habitat could help, the study determined, but too thick a shielding might also increase the dose of secondary radiation, which results when cosmic rays strike the craft and generate rebounding particles. Given that, any discussions of sending crews with today's tech very much seemed like a suicide mission.

Then again, encouraging advancements were also coming online. Beginning in 2021, a microwave-oven-sized device aboard the Mars *Perseverance* rover had been able to cleave oxygen molecules for potential use in both breathable air and rocket propellant. The machine, known as MOXIE (Mars Oxygen In-Situ Resource Utilization Experiment), devised at the Massachusetts Institute of Technology, was able to generate oxygen sixteen times.

"We are going to have to learn to be less Earth-dependent," explained Niki Werkheiser, director for technology maturation in NASA's Space Technology Mission Directorate. "That means being able to use the local resources in whatever ways make sense. This could include consumables for the crew, such as oxygen and water, propellant for fuels, or materials for construction of things such as landing pads, roads and potentially even habitats. Longer-duration missions further from Earth become much more feasible when we can use those local resources to help us live and operate so far from home."

Even with viable supply chains, however, establishing a colony would still require living off the land. And so, it turns out, that cricket bar I was handed wasn't that far off the mark. In fact, in a study titled "Feeding One Million People on Mars" and published in the academic journal *New Space,* Kevin Cannon and Daniel Britt made the case for crunching bugs as an almost imperative for any human settlement:

> Here, we consider the more radical goal of producing enough food on Mars to sustain a permanent settlement of private citizens that increases to 1 million people within 100 Earth years. We modeled a population that grows from immigration as well as naturally. Calorie needs were calculated on a per-person basis, and land use was modeled with a diet that includes staple crops, insect products, and cellular agriculture. Food self-sufficiency can be attained within 100 years with reasonable inputs, but massive amounts of imported food would be needed in the interim.

For argument's sake, if humans could indeed develop sustainable Martian settlements, there were also other, equally foundational questions that would have to be asked. How might they be governed? In what ways would international competition manifest, particularly with China? And how would challenges like resource scarcity and those profound psychological effects of isolation shape social dynamics and structure?

If we look deep enough, history might actually offer us a few clues.

· CHAPTER 33 ·

# GOVERNING MARS

The year was 1606. A group of English merchants and nobles had decided they needed three transatlantic ships to pursue riches in "the New World." Rivals Spain, France, and Portugal had already amassed immense wealth across Latin America. England—under King James I—was looking to catch up. The Crown, however, was also wary of risking public money. So to finance the effort, the men formed what became known as the Virginia Company, a joint-stock group that allowed investors to pool their money and fund voyages to a narrow peninsula with a freshwater source along the Chesapeake Bay. King James granted the company a royal charter, giving it the power to devise its own leadership and establish England's first permanent colony in the New World. Governance would come in the form of a corporate entity, with early colonists effectively employees. While a local council would be formed and wield some authority over local matters, overarching control rested with the company in London, whose influence touched virtually every aspect of colonial life.

That suggests, if history is any guide, this may also be how the first Martian settlements are established, given it is quite plausible the commercial sector will again play an outsized role; something those at SpaceX were clearly contemplating. Consider, for instance, the promises of future off-Earth markets during early Starlink beta testing, which drew comparisons with that early Virginia Company. Buried in the beta software agreement of SpaceX's Starlink app were the following sentences:

> For Services provided on Mars, or in transit to Mars via Starship or other colonization spacecraft, the parties recognize Mars as a free planet and that no Earth-based government has authority or sovereignty over Martian activities. Accordingly, Disputes will be settled through self-governing principles, established in good faith, at the time of Martian settlement.

Published in 2020, it might have at first seemed like a joke, given there are certainly no Starlink satellites orbiting Mars. (Musk is also somewhat known for his gags.) And yet in an interview with Law360 that same year, SpaceX's then–general counsel, David Anderman—who could not be reached for comment to independently verify these statements—seemed to double down, reportedly saying he was drafting a constitution for Mars.

"I think SpaceX will move to impose our own legal regime," he reportedly told the New York–based legal news service. "I think it will be interesting to see how it plays out with terrestrial governments exerting control. I do think we are going to have a pretty important role to play in what works and what laws apply."

Neither Anderman nor current leadership at SpaceX was available to provide further comment or context to those statements. But the ideas predictably spawned a firestorm of controversy. In an article published in *SpaceNews,* titled "No, Mars Is Not a Free Planet, No Matter What SpaceX Says," Italian space lawyer Antonino Salmeri argued that "refusing any Earth-based authority over Martian activities conflicts with the international obligations of the United States under the Outer Space Treaty, which naturally take precedence over contractual terms of services."

In fact, Article VIII of the Outer Space Treaty says that states are to "retain jurisdiction and control" of registered space objects and "any personnel thereof, while in outer space or on a celestial body." It relies on a concept known as quasi-territorial jurisdiction, whereby a state continues to maintain its authority over ships, aircraft, and spacecraft, even while operating in far-flung autonomous or semiautonomous regions. If SpaceX decided to start launching from non-treaty-member countries, the craft would thus still remain bound under international law by the company's home country.

"The bottom line is, if Musk wants to go to Mars and establish a settlement, the U.S. government's going to get involved," explained Michael J. Listner, attorney and founder of Space Law and Policy Solutions. Under President Trump, however, these reins had been significantly loosened. In the wilderness of Mars, it is easy to see how those early outposts might resemble far earlier company towns that once sprang up across America's wild frontiers, where coal miners, lumberjacks, and railroad workers all eked out a living with limited oversight beyond the company itself. In those isolated locales, often devoid of external competition, workers frequently faced steep housing costs, low wages, and looming debts that had to be worked off before they could leave. Still, unlike on Mars, leaving *was* an option.

On the Red Planet, it likely wouldn't be. And here again, we see the value of precedent.

In 2012, for instance, a Dutch nonprofit announced its Mars One project to send four astronauts on a one-way trip to the Red Planet. Never to return to Earth, they were to build humanity's first permanent settlement. Yet the group was roundly criticized for its lack of logistics and medical concerns, as well as critical gaps in hardware planning. It eventually declared bankruptcy. But news coverage of the mission prompted deeper questions about the nature and perils of such extreme isolation, following months of travel, regardless of the arrangements at the outset of the journey.

Once again, the parallels to Jamestown are striking.

Back then, getting to the "New World" involved a perilous, months-long voyage, fraught with high mortality rates, limited navigational tools, and countless unknown dangers. Inevitably perhaps, those realities set the stage for tensions between three groups: (1) the early settlers, principally concerned with establishing a defensible, self-sustaining colony; (2) distant investors, eager for a return on their investments; and (3) the English crown, which tended to view Jamestown as a more long-term extension of its imperial influence, as well as a newfound source of economic gain.

One would think, therefore, a further blending of national interests and private enterprise would be likely in space, and on Mars. And given that Musk says he expects one million people to live there in a couple decades (a figure that even the most optimistic of Mars observers openly doubt), it makes sense to discuss what those societies might start to resemble.

If Martian settlements grow and evolve, it is natural to assume they will do so with varying and competing social and legal structures. A SpaceX settlement, for instance, would tend to look quite different from one devised by authorities in Beijing. Martian immigrants could find themselves the focus of bidding wars between colonies, shaping the future of Martian society as competition for talent and labor develops, and notions of a what-do-you-bring-to-the-table-type ethos emerge. It is often on those perilous frontiers, where survival is far from certain, in which practical skills matter more than background. Perhaps, in its most optimistic viewing, that could have a greater equalizing effect on social strata than what we experience. Chances are, however, the haves-and-have-nots structure we have on Earth would merely be exported. The dichotomy is perhaps aptly captured in the 2016 science fiction film *Passengers,* in which Chris Pratt's character as a mechanic is given a preferential spot aboard an interstellar spacecraft heading to the planet Homestead II, albeit accruing heavy debt along the way. In an example of how fiction sometimes precedes reality, Musk even said in a January 2020 Twitter thread that Mars "needs to be such that anyone can go if they want, with loans available for those who don't have money." His comments, however, drew comparisons to early indentured servitude, in which settlers signed labor contracts without salaries (or for minimal compensation) to repay debts they accrued to gain passage to those early American colonies. The first such servants began arriving in Jamestown in 1607, filling a need for cheap labor, in a system developed by the Virginia Company. Workers were typically bound by four-to-seven-year contracts, in exchange for a spot aboard the ship, as well as food and lodging upon arrival.

Conditions were harsh, and servants could be sold or lent out to other colonists or distant corporate managers for the time of their contracts. Those agreements could also be extended as punishment for attempting to break the contract by running away.

A growing reliance on enslaved Africans would eventually render the use of indentured servitude obsolete, while nineteenth-century legal reforms would later undercut the practice, even though it would reemerge with Asian immigrants throughout the Caribbean and South America. Today, debt bondage and human trafficking remain prevalent in many parts of the world—a kind of indentured servitude that has proved difficult to police on Earth, let alone 140 million miles away. It seems, frankly, almost unknowable how such legal structures, imbued with the values we ascribe to them, might be practically applied to Martian colonies, millions of miles away.

Perhaps an earthly legal overseer would need to be included in those early colonies to keep them in line, reporting back to regulatory bodies, despite a wide range of delays in one-way communications from Mars to Earth. Still, how then could that observer remain objective when the very structures of his or her survival would almost inevitably rely on the very companies that they reported to? Meanwhile, for those concerned about worker conditions, Musk seemed to be doing little to assuage their concerns. His praise of Chinese workers (who are not subject to the kinds of labor protections that benefit their Western counterparts) for "burning the 3 am oil" seemed to raise even more questions about what kind of environment might be in store for future Martian workers.

"[Chinese workers] won't even leave the factory type of thing, whereas in America people are trying to avoid going to work at all," Musk reportedly explained in praise of the workforce at Tesla's facilities in Shanghai, just as unsubstantiated reports surfaced of Chinese workers pulling twelve-hour shifts, six days per week, while earning less than $1,500 per month. Those reports could not be independently verified. Still, given that Mars would require a cross section of workers (and robots) to thrive, competition between settlements seemed not only possible but likely. For those like Mars Society president Robert Zubrin, that competition might also help sidestep potential perils of a more mercantile form of governance.

"The ones that have the best ideas and create the most attractive forms of society are the ones that will draw the most immigrants," Zubrin told me over the phone.

It's certainly possible. Of course, again if the past is any teacher, new world competition also breeds inequality, marginalization, and conflict, which Zubrin acknowledged. The challenges of living on such a foreboding planet as Mars, where residents would face radiation exposure, limited gravity, toxic dust, punishing temperatures, and extreme isolation and confinement, could also exacerbate these divides and lead to a far more dystopian reality than many who dream of a multiplanetary existence readily accept.

As enticing as it might sound to work around the clock in a corporate-run cave on a planet where the outside environment can kill or progressively poison humans, reservoirs of water, carbon, nitrogen, hydrogen, and oxygen were indeed there. And they could be the building blocks of a more advanced civilization. There is little debate that a human settlement on Mars would be a pivotal moment in the history of our species, with all the downstream technological and exploratory benefits that could pave the way for a truly spacefaring civilization. As Thomas Paine once wrote, referencing the uncertain emergence of his new nation in 1776, we would "have it in our power to begin the world over again."

But to make it work, that colony would have to be self-sustaining. If supply ships from Earth stopped coming, could it actually survive? Affirming that possibility implies the existence of an immensely complex and resilient system, which is precisely the kind of challenge referenced in something called the great filter theory, which seeks to explain why intelligent life in the universe appears to be so exceedingly rare. Forged by the work of an Italian-born physicist, it's commonly a framework when considering interplanetary habitats and necessary precursors to the prospects of greater human longevity across the cosmos.

And in many ways, it's a conversation that began in earnest over lunch.

· CHAPTER 34 ·

# WHERE IS EVERYBODY?

Enrico Fermi had just sat down for a meal. It was the summer of 1950, and the Nobel Prize–winning physicist, who had worked on the Manhattan Project, was joined by a few colleagues to discuss the possibility of intelligent life on other planets. He then asked a simple question: Where is everybody? If the cosmos were indeed filled with advanced alien civilizations, Fermi posited, then surely detectable evidence of that life would be scattered across the universe. Given the breadth of human history, one might expect the question had been asked before. But during Fermi's era—one that coincided with revolutions in nuclear power and physics and thus offered new and unprecedented potential for advanced space travel—the question somehow seemed more urgent. The Fermi paradox, as it later became known, has since been roundly debated across research labs and universities, given the apparent contradiction between the high probability of extraterrestrial life and a sheer lack of evidence, so much so that an organization dedicated to the search for intelligent life, better known by its acronym, SETI (or Search for Extraterrestrial Intelligence Institute), sought to elaborate on that observation:

> Fermi grasped that any civilization with a modest amount of rocket technology and an immodest amount of imperial incentive could rapidly colonize the entire Galaxy. Within a few tens of millions of years, every star system could be brought under the wing of empire. Tens of millions of years may sound like a long project, but in fact it's quite short compared to the age

of the Galaxy, which is roughly a thousand times more. So what Fermi immediately recognized was that the aliens have had more than enough time to pepper the Galaxy with their presence. But looking around, we don't see any clear indication that they're out and about. We don't see any obvious evidence of a galactic empire or a United Federation of Planets.

Enter Frank Drake.

In 1961, the Harvard-trained radio astronomer, who would later serve as chairman of the SETI Institute's board of directors, endeavored to help quantify the odds. Convinced that the search for intelligent life would eventually bear results, he devised a probability formula that employed estimates for several variables, including the rate of star formation in the Milky Way galaxy, as well as the ratio of stars to planets within the so-called habitable zone where liquid water could exist. It also included an estimate of the fraction of those planets where intelligent life could develop, capable of producing discernible communications, not to mention the length of time those civilizations would need to send their signals out into the cosmos. The challenge, however, lay in the fact that such estimates were, at least in part, based on data sets that were fundamentally unknowable. Still, Drake wrote in his 1962 book, *Intelligent Life in Space,* that with "almost absolute certainty, radio waves sent forth by other intelligent civilizations are falling on the earth."

"A telescope can be built that, pointed in the right place and tuned to the right frequency, could discover these waves," he added. "Someday, from somewhere out among the stars, will come the answers to many of the oldest, most important and most exciting questions mankind has asked."

It was a big claim, and one that presupposed a number of factors, including that radio waves would actually be how an advanced alien civilization might communicate. In 2020, the physicist Marek Abramowicz from the University of Gothenburg, whose research centered on black holes and quantum gravity, would propose an alternative. Referred to as the "galactic centre gravitational-wave messenger," the idea centered on the detection and study of gravitational waves, and whether gravity itself might be a more likely beacon. In a research paper published in the journal *Scientific Reports,* the study concluded that "all technologically advanced species must be aware of the unique property of the galactic centre: it hosts Sagittarius A* (Sgr A*), the closest supermassive black hole to anyone in the Galaxy." Sagittarius A is located at the galactic center of the Milky Way, with a mass more than four million times that of our Sun. The paper then noted that "a civilization with sufficient technical know-how may have placed material in orbit around Sgr A* for research, energy extraction, and communication purposes." It then went further to suggest that "placing and sustaining a device near Sgr A* is therefore astrophysically possible. Such a probe will emit an unambiguously artificial continuous gravitational wave signal that is observable with LISA-type detectors." The paper's reference to LISA, or Laser Interferometer Space Antenna, referenced a planned mission by the European Space Agency for 2035 for humanity's first space-based gravitational wave observatory.

Gravitational wave detectors would be comprised of three spacecraft 2.5 million kilometers apart, designed to detect ripples in space-time caused by the mergers of black holes. The idea would be to develop insights into both our galactic evolution and black hole formation, among other areas of astrophysical research. This new way of sensing cosmic events could unveil phenomena humans have never before detected, raising even more questions and speculations as to what we might find. Could such a system, for instance, be used to suss out the signals of distant civilizations? It seemed the stuff of science fiction. Yet there was serious scholarship being carried out on precisely those subjects, not to mention a precedent for similar instruments making remarkable discoveries. In 2015, for instance, the Earth-based LIGO, or Laser Interferometer Gravitational-Wave Observatory, had already etched itself into the history books by detecting the first gravitational waves following the collision of two black holes. When it came to a space-based system, what else might be found?

Abramowicz, however, wasn't quite finished speculating, referencing humanity's first interstellar probes, Voyager 1 and Voyager 2, and what precedents they might ultimately represent.

"The Voyagers carry not only instruments designed to study the planets in the outer Solar System, but also two golden records with messages intended for future humans or extraterrestrial lifeforms," the paper noted. "Chances are that other civilisations would also try to announce their presence to the Galaxy. Using probes with instruments that emit radio waves aboard is one of the most obvious strategies to us. Indeed SETI, our most elaborated search programme for extraterrestrial intelligence, is based primarily on analysing radio waves. . . . Of course, there is no particular reason to limit searching and announcing activities to radio waves."

Yet those like renowned cosmologist and astrophysicist Stephen Hawking favored a more cautious approach, considering the practice of sending signals out into the universe in the hopes of contacting other intelligent life too risky for humanity. "If aliens visit us, the outcome would be much as when Columbus landed in America, which didn't turn out well for the Native Americans," he said during a 2010 Discovery documentary. "We only have to look at ourselves to see how intelligent life might develop into something we wouldn't want to meet."

Of course, signs of life could also take other forms.

Those like the theoretical physicist Avi Loeb at Harvard University are often quick to point out that the search for intelligent life need not be for living organisms, given that the cosmic timescales of a 13.8-billion-year-old universe provided ample room for civilizations to surface and go extinct well outside the scope of mere human existence. Scouring the cosmos for signs of existing alien life would presumably dramatically reduce the chances of uncovering evidence of life elsewhere. Loeb, who controversially came to global prominence when he suggested a hundreds-meter-long interstellar object that passed through Earth's solar system in 2017 could have been an interstellar probe, has devoted much of his career to the study of space archaeology, or the search for historical artifacts that might offer clues about life beyond Earth.

The Drake equation "misses a crucial possibility: most technological civilizations that ever existed might be dead by now," Loeb wrote in a 2019 Harvard paper. "There are two obvious reasons to suspect that this might indeed be the case," he continued. "First, as soon as we mastered advanced technologies, we also developed the means for our own destruction through catastrophic nuclear, biological or chemical wars, or through a global change in our habitat. Second, recent data from the Kepler satellite implies that about a quarter of all stars host a habitable, Earth-like planet. This naturally reinforces a paradox, formulated in 1950 by the physicist Enrico Fermi. . . . But this does not mean that we cannot prove other civilizations existed. On Earth, we find evidence for ancient cultures that are not around anymore, like the Mayans, through the artifacts they left behind. Similar to the work of archaeologists who dig into the ground, astronomers can search for technological civilizations by digging into space."

It was a fair point. But the Drake equation was also devoid of another key component: the confirmation of exoplanets, which are defined as any planet that exists beyond our solar system. While astronomers speculated about more vast scatterings of distant planets, the first one would not be confirmed until 1995, more than three decades after Drake first publicly posed that dichotomy. Astronomy, to be sure, has evolved a long way since the 1960s. The Very Long Baseline Array, for instance, which includes a series of ten radio telescopes spread across five thousand miles, as well as the launch of the Hubble Space Telescope in 1990 and China's Five-Hundred-Meter Aperture Spherical Telescope located in Guizhou province, had not existed at the time Drake first devised his equation. Telescopes at large were becoming exponentially more formidable, having already vastly expanded our observations in ways scarcely imaginable in Drake's time.

Coincidentally, it would be the man who ran NASA beginning in 1961—at the start of the first space race—and whom the agency credits with doing "more for science than perhaps any other government official," who would have his name attached to what would soon become the largest and most powerful telescope in space—ushering in an almost philosophically new and Copernican-like awareness of our place in the cosmos.

His name was James Edwin Webb.

· CHAPTER 35 ·

# THE SEARCH

In the world of scientific soft power and cosmological preeminence, there are few more illustrative examples than the images captured by the James Webb Space Telescope. Run mostly by NASA's Goddard Space Flight Center, along with the European Space Agency and the Canadian Space Agency—and named after the NASA administrator during those early Apollo years—the telescope captures enchanting signatures of infrared clouds of distant nebulae and spiral galaxies. So much so that Chinese scientists have pursued their own plans for a rival observatory, known simply as the China Space Station Telescope (CSST), albeit with wider, Hubble-like optics, meant to operate alongside the country's Tiangong station in low Earth orbit. With a field of view some three hundred times larger than that of Hubble, CSST is expected to be used in the deciphering of dark matter and other enigmas of the early universe.

This, perhaps, is one of the clearest advantages yet of the kind of geopolitical competition that can prompt investment, innovations, and deep space ambitions, prompted by two rivals pushing each other deeper into the cosmos and toward a moment that marked perhaps the very inception of space and time as we know it.

In the years that followed the Big Bang, the first stages of our cosmic evolution would give way to a primordial soup of light and particles. As the universe cooled, electrons combined with nuclei. And over time, gravity attracted more material, ultimately forming dense cores. Inside, temperatures swelled and nuclear fusion began, thus forming the first stars, and eventually first galaxies, a period known as the Cosmic Dawn. These faint stretches of light and matter are precisely the sort of phenomenon that such telescopes were designed to explore. And because it takes millions or even billions of years for light to reach Earth, gazing out at such distant galaxies is like peering back in time. Among those findings is the most distant galaxy ever seen, JADES-GS-z14-0, which corresponds to a time less than 290 million years after the universe itself was formed.

Humans, in fact, never had the ability to gaze so far and with so much precision until after an Ariane 5 rocket launched from Europe's spaceport in French Guiana on Christmas Day 2021, delivering James Webb into a special orbit, roughly a million miles from Earth. Unlike the Hubble Space Telescope, which orbits Earth, the James Webb Space Telescope orbits the Sun, playing detective to faraway worlds that never quite seemed so clear. Beaming back more than 57 gigabytes of data daily, the telescope, however, also reveals information closer to home—an underground ocean on one of Uranus's icy moons, as well as a molecule called dimethyl sulfide on a planet 120 light-years away, which is usually associated with life. It also uncovered evidence of carbon dioxide across the icy surface of Europa, one of Jupiter's moons, massive plumes of water vapor from Saturn's moon Enceladus, and distant nebulae home to stars more than a hundred times the size of our Sun. A never-before-seen molecule in space known as methyl cation was also detected roughly 1,350 light-years away in the Orion Nebula, which NASA identifies as the very cornerstone of interstellar organic chemistry.

To see such stunning and unprecedented high-resolution images of the universe's distant parts also engenders another phenomenon, comparable to the Apollo Moon landings, given the sheer awe and wonder they evoked in the public consciousness, as well as the downstream effects they can bring. Each seemed to help reinvigorate broader interest in space science and the pursuit of distant worlds and questions of life in the universe, even as its operators worried about the telescope's longevity and continued sources of funding. That was especially true in 2025, amid questions of budget cuts to NASA's science funding, imperiling James Webb and Hubble at the precise moment when Chinese deep space efforts were racing to catch up.

Nestled in the foothills of the Andes mountain range, not far from an Argentine army garrison, is China National Space Administration's Espacio Lejano deep space station. In operation since 2018, it was considered instrumental in providing support for the first-ever far side of the Moon landing, carried out by China's Chang'e 4 mission.

CERN, the world's largest particle physics laboratory, near Geneva on the Franco-Swiss border, July 2022. Physicists found evidence in the Large Hadron Collider data of a particle consistent with the Higgs boson, a quantum field that fills the universe and gives physical objects their mass.

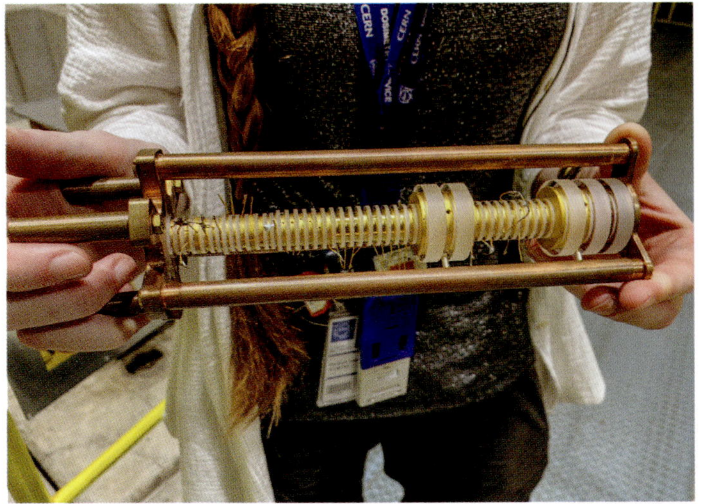

Holding an antimatter trap, or Penning trap, able to receive and release antiprotons produced at CERN's Antiproton Decelerator, CERN's Antimatter Factory, Prévessin-Moëns, France, along the Swiss border, July 2022.

Inside CERN's Neutrino Platform, which houses a prototype of the Deep Underground Neutrino Experiment, known as ProtoDUNE. The facility is filled with argon, and is used in advanced neutrino research. The goal is to investigate some of the biggest mysteries in particle physics, including how neutrinos have factored in the universe's evolution.

Countdown clock prior to the launch of IM-1, February 2024

A tense mission control where the Intuitive Machines team in Houston, Texas, prepares for *Odie's* touchdown on the Moon.

This 10.5-billion-year-old star cluster, NGC 6496, is situated roughly 35,000 light-years away from Earth, with brightness that changes over time that gives astronomers a trove of information, including stars' sizes, temperatures, brightnesses, chemical makeups, and even how they may have evolved over time.

(*Left*) The first-stage engines of a Rocket Lab Electron rocket ignite as the vehicle blasts off from Launch Complex 1, Pad B, in Māhia, New Zealand, carrying two NASA CubeSats designed to study tropical cyclones, hurricanes, and typhoons.

Micaela Landivar, Intuitive Machines radio frequency communications engineer, during a landing simulation to prepare for IM-1, 2023.

Lieutenant General John Shaw served as the deputy commander of the U.S. Space Command from 2020 to 2023. During the Ascend conference in Las Vegas, Nevada, Shaw reflected on the intersection of commercial sector and national interests and the growing security needs in space.

The IM-2 mission lunar lander, *Athena,* took this image during low lunar orbit approximately twenty-four hours before touching down on the lunar surface on March 6, 2025.

Following an analogue rover exercise outside the Mars Desert Research Station in Utah, the author, David Ariosto *(at far left),* is with *(clockwise from bottom left)* Nadia Maarouf, clinical scientist at the University of Calgary; *(third from left)* Luca Rossettini, D-Orbit's CEO and founder; *(fourth from left)* Vittorio Netti, senior projects engineer at Axiom Space, in 2022.

To that point, it's worth considering James Webb and similar technologies not merely with a scientific lens but also through a geopolitical and even economic one, given its discoveries are not made in a vacuum, and thus add to a frontier of knowledge that often eventually become useful in both the national and the commercial space, even if that sort of value is not usually immediately apparent. These sorts of investments, to be sure, can translate into long-term power and influence, something of which leadership in Beijing seems to be acutely aware.

Consider, for instance, James Webb's explorations of the metal-rich asteroid Psyche—situated in the main asteroid belt between Mars and Jupiter—which has excited scientists to learn more about the early formings of planets, as well as commercial entities interested in prospective mining of celestial bodies. Here again, in fact, we see China's ventures stand out. At China's University of Mining and Technology, a spider-looking six-legged machine designed by a research team led by Dr. Liu Xinhua by 2025 was in the works as a prototype for the kind of deep space mining that might one day be employed on an asteroid like Psyche.

"This claw system is an array-type structure that enhances adhesion and gripping ability in microgravity environments," Liu noted. "It allows the robot to stay anchored while collecting samples and move efficiently based on the terrain."

Almost simultaneously, in what seemed to be the stirrings of a space mining race, a California-based company called AstroForge sought to one-up their Chinese rivals, planning to hitch a ride aboard Intuitive Machines' upcoming mission and deliver into deep space a probe that could observe one of the company's target asteroids.

"Will we actually land on an asteroid and get these beautiful samples? Probably f—— not," AstroForge's CEO, Matt Gialich, told Payload. "But do we hope to change the name of the game for access to deep space, and show people that the price point we're doing this at is doable? I hope."

Sparked, in part, by discoveries made possible through James Webb, these seemed to be the first steps in exploring the commercial potential of the solar system's scattered space rocks, some containing water ice, iron, and nickel, which could one day prove valuable, particularly should in-space manufacturing ever become reality. Meanwhile, other asteroids containing platinum-group metals were thought to be attractive given their relatively low supply on Earth and range of uses.

"A single one-kilometer-diameter asteroid, if it was platinum-bearing, would contain about 117,000 tons of platinum," Mitch Hunter-Scullion, the founder and chief executive of the Asteroid Mining Corporation, postulated to *The New York Times*. There is, of course, the not-so-small matter of actually landing on an asteroid, retrieving its contents, and returning them to Earth with sufficient quantity that would offset those operational costs and eventually turn a profit.

But I digress.

Beyond pure profit, these asteroids also held a far darker significance: the looming threat of a catastrophic strike against the Earth, a silent menace with almost unimaginable destructive power. In fact, a series of close calls over the years had broadened the conversation, acting as reminders that these orbiting jackpots were also hurtling projectiles through space with trajectories that could one day rewrite life on our planet.

It, of course, had happened before.

But that also seemed like an opportunity.

Asteroid strikes were a global concern, not just a geopolitical one—a threat where borders and rivalries didn't matter. Perhaps then, danger to the collective of humanity could serve as a rallying point. Indeed, there were openings. In 2024, a man by the name of Zhou Guolin, then China's minister counselor for science and technology, said "history has proved that isolation is not a solution, and that cooperation is the only solution to go forward." In this burgeoning new space race, rife with rivalry and conflict, finding such areas of collaboration seemed more important than ever, especially when it came to things as important as planetary defense.

One such asteroid—ominously named after an ancient god of darkness and disorder—was, in fact, already hurtling toward Earth.

· CHAPTER 36 ·

# DEFENDING EARTH

In Egyptian mythology, Apophis is nemesis to the Sun god, Ra. A serpent-like deity linked to chaos and destruction, it is thought to be evil incarnate, seeking to plunge Earth into eternal darkness. Coincidentally, it is also the name given to a near-Earth object, nearly the length of the Empire State Building, and identified as "one of the most hazardous asteroids that could impact Earth."

At one time, it hurtled through space in the asteroid belt between Mars and Jupiter. But over time, largely due to the gravitational effects of Jupiter, it was ejected from that orbit and drawn into one decidedly similar to that of Earth. Given its size and composition, if Apophis ever struck our planet, the energy released would be tantamount to "tens to hundreds of nuclear weapons." That little tidbit is especially terrifying given that on June 19, 2004, it actually came into view of astronomers affixing their telescopes in the Sonoran Desert.

Composed of mostly silicate and a mixture of nickel and iron, this rotating, peanut-shaped object, also known as 99942, was visible only briefly due to weather conditions and a series of technical issues. But that was long enough to make the call. Working at the Kitt Peak National Observatory, situated in Arizona's Quinlan Mountains, astronomers Roy Tucker, David Tholen, and Fabrizio Bernardi alerted NASA's Center for Near Earth Object Studies, run by the agency's Jet Propulsion Laboratory, or JPL, in Pasadena.

Paul Chodas was working that day and received the alert. "We were anxious to get more data," he told me in a phone call from his office. "To either confirm that it's going to hit, or much more likely to rule out the chance."

"We run the numbers and get the probability," he added. "But getting the data is the key thing." Initially, his team determined that the chance of impact was about one in thirty-seven. No one seemed to like those odds, but at least they would have a three-decade window to come up with a plan. NASA headquarters was alerted. A year later, Congress would act, mandating the tracking and cataloging of at least 90 percent of all near-Earth objects 140 meters in size and bigger. Then, relief. Subsequent data for Apophis seemed to all but rule out an impact, at least in the next century, removing it from the agency's risk table, albeit with an important caveat: Earth's gravitational effects, or a nudge from another asteroid, could always affect its trajectory in unexpected ways.

Either way, humans should again be able to see it relatively soon, this time with just a naked eye.

On Friday, April 13, 2029—yes, Friday the thirteenth—Apophis is expected to pass unsettlingly close to our planet, a mere nineteen thousand miles at its nearest point at an altitude that is actually below many of Earth's geostationary satellites. In fact, it may still have an infinitesimally small chance of striking Earth, according to an August 2024 paper published by the astronomer Paul Wiegert in *The Planetary Science Journal.*

"Our uncertainties are somewhat larger than JPL's but sufficiently accurate for the sensitivity study presented in this paper," Wiegert noted.

"From our simulations, we find that a $\Delta v$ of $3 \times 10^{-4}$ m $s^{-1}$ is sufficient to move Apophis 200 km on the *b*-plane to the nearest keyholes," he wrote. "Keyhole" is a reference to the trajectory an asteroid could take that would put it on a potential collision course with Earth. And while such an incident appeared highly unlikely, "monitoring Apophis telescopically for signs of such an event at the earliest possible date is recommended," Wiegert suggested.

Swiftness and federal action, however, rarely seem to go hand in hand. In fact, after Apophis's initial detection, it would take another twelve years before NASA's Planetary Defense Coordination Office could be established within the agency's Planetary Science Division of the Science Mission Directorate. In the interim, space objects would continue to strike our planet, and sometimes with major consequences.

In the air above Chelyabinsk, a western-central Russian city near the Ural Mountains, for instance, dashboard cameras would in 2013 capture footage of a fireball that scorched its way to Earth. The resulting shock wave left more than twelve hundred people injured. When I spoke with Paul, he actually described it as a "tiny asteroid," even though it weighed up to ten thousand metric tons, with a blast that damaged at least three thousand buildings and held the equivalent force of some 440 kilotons of TNT. Four days later, satellite imagery showed that its debris plume had already circumnavigated the globe. Three months later, that plume could still be detected.

This was "tiny."

It was, in fact, a much larger plume some sixty-six million years ago, derived from the impact of a nine-mile-wide asteroid that slammed into Mexico's Yucatán Peninsula, that many researchers believe brought an end to the age of dinosaurs, a near-apocalyptic event that wiped out roughly three-quarters of all life on Earth. Airborne soot blocked much of the Sun's light, spewing a sulfur aerosol into the atmosphere. Of course, there is an important distinction here, one often attributed to science fiction writer Larry Niven:

> The dinosaurs became extinct because they didn't have a space program. And if we become extinct because we don't have a space program, it'll serve us right!

Efforts to study the city-killing-sized asteroid were of course under way, including NASA's OSIRIS-REx mission—which itself sounds like a dinosaur—even as federal budget cuts threatened the program's longevity, with a University of Arizona–led project having been expected to survey Apophis during its earthly flyby in 2029. For all the monitoring, however, clear gaps in coverage would persist. Because asteroids are so often detected optically, the glare of the Sun can create significant blind spots in Earth's planetary defense—as evidenced by the Chelyabinsk event.

"[That meteor] was coming from the direction of the Sun," Paul explained.

To change that, NASA's Near-Earth Object (NEO) Surveyor, initially planned for 2027, was to be the first space infrared telescope built expressly to detect both hazardous asteroids and comets (which usually approach faster and with less predictability than asteroids) that come within thirty million miles of Earth's orbit. It would essentially broaden the practice of mere visual detection and provide for a more comprehensive accounting of potential threats. And yet its price tag had been growing. What in 2019 had been projected to cost as much as $600 million has since ballooned to $1.2 billion, and as of the writing of this book seemed especially vulnerable to the chopping block.

"This is a problem we know how to solve," explained Lori Glaze, NASA's Planetary Science Division director, in defense of the program. "How many global problems do we know how to solve that are of the order of $1 billion?"

And yet even *if* a planet-killing asteroid were discovered to be on a collision course with Earth with sufficient time to act, humanity would still need a plan.

· CHAPTER 37 ·

# IMPACT

On September 26, 2022, NASA's Double Asteroid Redirection Test (DART) mission struck the asteroid Dimorphos in the binary asteroid system Didymos. It represented the first time scientists had exhibited the potential to "fire back" at asteroids in a demonstration of deflecting them through impact, thus redirecting the trajectory of the asteroid. The graphic on this poster, just outside Johns Hopkins University Applied Physics Laboratory, illustrates the DART mission en route to Dimorphos.

When I first arrived at Johns Hopkins Applied Physics Laboratory (APL), Andy Cheng was "a nervous wreck." He was pacing outside a media-staging area at the Maryland campus in a scene that seemed torn from the pages of the latest Bruce Willis box office thriller. In reality, he was just taking a break from answering reporter questions about whether scientists could pull off an asteroid strike seven million miles away.

No big deal.

Over the years, Andy and his team had designed and managed a refrigerator-sized spacecraft that was now traveling at speeds of roughly four miles per second en route to a 492-foot asteroid called Dimorphos. They, being the APL team for NASA's Planetary Defense Coordination Office, wanted to see if they could strike and redirect an asteroid that itself was traveling about fourteen thousand miles per hour, and located more than twenty times the distance between Earth and the Moon.

It seemed like shooting a bullet with a BB gun at impossible distances. And Andy, who was not only chief scientist but also the man who helped come up with the idea of what would eventually be known as the Double Asteroid Redirection Test (DART), was now visibly feeling the gravity of the moment.

Impact, he hoped, was just a few hours away.

"I don't have any objective reason to be afraid," he told me, while waiting outside the facility. "There's nothing I know of that is going to make us miss, or that we haven't done. On the other hand . . . it's just this feeling."

Having launched DART in November 2021 with the intention of targeting a binary asteroid system, scientists felt they could better determine if the target's trajectory had indeed been altered, given that the orbit of the smaller asteroid would presumably be slowed by a direct hit. Dimorphos posed no danger to Earth. But the idea was that if a spacecraft could even slightly alter an asteroid's trajectory, a similar approach could be used against a planet killer, or other, less severe asteroids. Even a small modification over time would represent massive trajectory changes given the expansiveness of space. Find an Earth-bound asteroid soon enough, the thinking went, and even the slightest nudge would be enough to move it off course.

Given it was NASA's first real foray into planetary defense, Andy had been there all morning. His team had developed what was called the Small-Body Maneuvering Autonomous Real Time Navigation system, which essentially enabled the craft to make its own autonomous guidance decisions as it approached its target. The technology was supposed to be able to distinguish between the two asteroids and then adjust trajectory as it got closer.

Now, the craft was on its final approach. Real-time four-pixel imagery was being beamed back to Earth and into the control room, which was on Johns Hopkins's big screen in full view of reporters. DART was only sixty-one thousand miles from its target. Scatters of dim white dots filled the screen, along with an ever-diminishing margin of error. At this distance, APL noted, even a "tiny maneuvering error could be the difference between hitting Dimorphos and racing past it at more than 13,000 miles per hour."

In many ways, Andy had spent his entire professional life geared toward this kind of moment. He had studied physics at both Princeton and Columbia Universities, earning his PhD at the latter, and had endeavored to reach an asteroid before, working as a project scientist on NASA's Near Earth Asteroid Rendezvous mission, the first spacecraft to orbit and, in fact, land on an asteroid.

"We were not even allowed to say that we were going to land," he recalled. "Because they [NASA] were afraid that it was going to fail.

"But it went down and landed," he said, with a smile. In fact, "the spacecraft continued to talk to us, and we were able to stay down on the surface for another two weeks and make some very nice measurements of the composition." Later, Andy would work on the Japanese-led MUSES-C asteroid mission—the first spacecraft to ever take samples from an asteroid as well as successfully land on and take off from an asteroid.

But DART was different.

It wasn't purely research. It was the first time humanity had endeavored to strike back at a phenomenon that once ended much of life on Earth. Public attention was now focused on the mission, as throngs of mainstream news reporters piled in for the event. Given the $324.5 million spent on the DART project, "it would be very bad for us not to succeed," Andy explained. Plus, a failure could have direct effects on the pricier spaceborne infrared telescopes in development, like the NEO Surveyor, considered the "next mission in the Planetary Defense program after DART." First, however,

DART would have to prove its worth without eroding public confidence that could undermine future missions.

And by now, the craft was edging perilously closer to Dimorphos. On-screen, what had been a collection of dimly lit white dots had given way to a single, blurry white blob. It was still faint. But with each passing second, the asteroid was getting clearer and closer. Meanwhile, what had been a tense scene at APL began to feel more like a tailgate party as the trajectory was more and more obviously aligned with the hulking asteroid. Dimorphos's craters and crevices were now becoming more distinct.

"C'mon, baby!" a voice shouted out in the crowd as images of a powder-white asteroid filled the screen. Individual bits of rock were now in full view. The blackness of space had all been cropped out of the live camera viewfinder as the white of the asteroid filled the screen. Then the transmission abruptly ended, the viewing screen turning crimson red.

Impact.

Sixty-six million years after an asteroid killed the dinosaurs, Earth had finally acquired the technology it needed to fire back. And it did so with a 1,280-pound spacecraft that struck Dimorphos at 7:14 p.m. EDT on September 26, 2022. The collision's impact was the equivalent of three tons of TNT. More important, it slowed the asteroid's orbit by more than half an hour. By bouncing radio waves and measuring the light curvatures of both asteroids, scientists at JPL would later conclude the physical shape of the object had also changed. That was the big concern. If the object wasn't sufficiently dense, or was composed of only a loosely cobbled surface, would the craft's impact still create a big enough change in its trajectory? Or would it be like shooting an arrow through a Jell-O mold? Initially, it wasn't clear.

"When DART made impact, things got very interesting," said Shantanu Naidu, a navigation engineer at NASA's Jet Propulsion Laboratory, who would lead a subsequent DART study. "Dimorphos' orbit is no longer circular: Its orbital period is now 33 minutes and 15 seconds shorter. And the entire shape of the asteroid

Scientists discuss with journalists, September 2022, NASA's Double Asteroid Redirection Test mission, a successful attempt to strike the asteroid Dimorphos and knock it off its trajectory.

has changed, from a relatively symmetrical object to a 'triaxial ellipsoid'—something more like an oblong watermelon."

A subsequent study also raised the possibility that tiny fragments of Dimorphos, which were ejected upon impact, could now be headed to both Mars and Earth, creating the first-ever human-made meteor shower. Those bits would be expected to burn up in the atmosphere. But generally, the mission was considered a triumph, the beginnings of a newly forged arsenal against dangerous space rocks.

And yet questions about the realities of operating in deep space were mounting, along with fresh curiosities about what humanity was truly capable of, especially by those voices from within NASA's more forward-leaning outposts, who often viewed asteroid detection not just as precaution, but as prelude. To them, humanity needed to be thinking well beyond our own solar system.

# PART III

# DEEP SPACE

· CHAPTER 38 ·

# ALPHA CENTAURI AWAITS

As a kid, Pete Worden was bored with the solar system. From his earliest memories, he recalled two children's books that illustrated what he meant. One was titled *Planets*. The other, *Stars*. The former had pictures of rocks and clouds, which he called "boring." The latter showed alien life, with tantalizing captions that read something like this: "Although we don't know of them in any planets or any other stars, we assume there are some."

From that moment on, Pete was hooked.

"Why don't they look for aliens?" he asked his mother.

People do, she replied. "They're called astronomers."

Without knowing it, his mom had set in motion a career that would lead Peter to become director of NASA's Ames Research Center and eventually executive director of Breakthrough Starshot, an ambitious initiative focused on engineering a proof of concept for a fleet of interstellar probes capable of reaching the Alpha Centauri star system.

At more than four light-years away, it is a distance that's difficult to wrap one's mind around. But for context, consider Voyager 1, an interstellar probe launched in 1977, just one of two spacecraft ever to journey beyond our solar system. After speeding away from our planet for nearly half a century at roughly 38,000 miles per hour, the craft has just about traveled one light-*day* away from Earth. Again, Alpha Centauri is at a distance of 4.3 light-*years,* or roughly 25 trillion miles. And that is the very *closest* star system to our Sun and just one of hundreds of billions in the Milky Way galaxy alone. Put another way, consider the fastest human-made machine ever built. Launched in 2018 to make observations of the Sun's outer corona, NASA's Parker Solar Probe travels at a blazingly fast 430,000 miles per hour. If that craft were redirected toward Alpha Centauri, it might still take somewhere around 7,000 years to make the journey.

Interstellar distances are, in other words, mind-bogglingly huge.

But back to our nearest neighbor, Alpha Centauri.

In August 2016, one of the world's most powerful planet hunters, a 3.6-meter visible-light spectrograph in Chile, discovered a new Earth-like planet in that system. By examining the so-called tug-of-war of gravity between a star and its orbiting planets, which creates a shift of the star's spectrum and tiny wobbles in a star's motion, a group of European astronomers found a rocky world only a bit larger than Earth. They called it Proxima b, which orbits the habitable zone around Proxima Centauri, a red dwarf star in the Alpha Centauri star system. It was among the more promising candidates suitable for life.

Four years later, in 2020, the team seemed to be on the brink of a remarkable discovery. Working on an initiative known as Breakthrough Listen, both a SETI project and a $100 million program established by Yuri Milner and Stephen Hawking, which employs an array of telescopes to survey a million nearby stars, the team uncovered what appeared to be radio signals. Subsequent analysis concluded that the signal was not in fact an "extraterrestrial technosignature," but rather electronic noise that resulted from radio interference of human technologies. Still, it was "undoubtedly one of the most intriguing signals we've seen to date," explained Andrew Siemion, who helped lead the effort. The detection stirred imaginations, while also raising harder questions about what life, if it exists elsewhere, might look like. If it were to exist in a place like Proxima b, it would likely face a host of challenges unfamiliar to the ways in which humans developed on Earth. First off, let's consider where it might exist. Proxima b, because it is believed to be tidally locked, meaning the planet does not experience the same day and night cycles as Earth, so there is the distinct possibility of climatic extremes on either side of it. If life there were to exist under conditions similar to those on Earth, the more promising locales might then be along the transitional regions between the two hemispheres, where more temperate climates would presumably be more likely to occur. Merely residing within a habitable zone, however, is hardly sufficient for life. The Moon proves that. Also, because Proxima b orbits a red dwarf star, as opposed to a more powerful solar-type star like Earth's Sun, its habitable zone requires it to be closer to the star to receive a similar level of luminosity. That shorter distance might then suggest that the planet could be more susceptible to solar flares, which could then make life especially difficult, particularly without the benefit of the kind of atmosphere that sufficiently absorbs solar radiation, or a magnetosphere that helps to deflect charged particles. It is not clear whether, or if so to what extent, Proxima b has a suitable atmosphere or magnetic field. And given that uncertainty, "habitability" may be too strong a word. What's more, in a computer modeling study published a year after the planet's discovery, its lead author, Katherine Garcia-Sage, a scientist at NASA's Goddard Space Flight Center in Maryland, concluded that if Earth were to exist where Proxima b is located, its atmosphere likely would not survive.

Nevertheless, under Pete's guidance, plans were now being devised to go there.

But first, let's paint a better picture of who this man is.

Having earned a doctorate in astronomy before serving as a researcher at the University of Arizona, Pete had spent much of his career in the U.S. Air Force, focused on large space optics for both national security and scientific purposes, as well as near-Earth asteroids and solar activity in nearby stars. It was military focused. But there was a reason for that.

"The reason I stayed in the military was like Jesse James when they asked him why he robbed banks," he explained. "That's where the money is."

He wasn't wrong. Battle cries have a certain way of securing funding. Yet Pete hardly fit the stereotype. Working at the Sacramento Peak Observatory in New Mexico early in his career, "I had a beard until the colonel found out about it, and said he was coming out to visit the next day.

"My uniform looked like shit because I burned a hole in one of the shirts trying to iron it. And I had a jacket on over it so he couldn't see the burn hole," he added.

"The only single girl there was the librarian," whom Pete married before departing New Mexico just a few years later.

"I was there for a few years and really had no intention of staying in the air force. I really wanted to go to astronomy, but the air force moves around. So I got reassigned to Los Angeles to work on a highly classified program." Under President Reagan, Pete would then help develop the Strategic Defense Initiative, which would enjoy roughly $30 billion in federal funding over the next decade, in large part to neutralize the threat of intercontinental ballistic missiles. During that time, he also served under the U.S. Air Force general James Abrahamson, who had taken up the role as the Strategic Defense Initiative's first director.

"I think Abrahamson originally thought that he'd use me as an executive assistant. It's basically a bureaucratic job that I was terrible at." For instance, "I was sitting in a chair right behind him, and he had some people in the room. He then turned to me and asked, 'What's the colonel's first name?' And I said, 'Colonel.'

"He expected a military assistant to know all those details," Pete said, with a kind of shrug of his shoulders. "And I'm terrible at details."

Still, Pete attained the rank of brigadier general before heading up NASA's Ames Research Center in Mountain View, California, which came with an annual $600 million operating budget that focused on aeronautics research and exploration technologies, backed by a twenty-five-hundred-person staff. That staff, he added, was the real resource. And under his leadership, the team would support a bevy of projects, including NASA's Mars *Curiosity* rover, as well as an airborne infrared observatory known as SOFIA, which flew into the stratosphere and allowed astronomers to study the solar system in ways not possible with ground-based telescopes. The team also supported the Kepler space telescope, considered NASA's very first expressly planet-hunting mission, which determined during its nine years in operation the existence of "billions of hidden 'exoplanets,' many of which could be promising places for life."

The hunt for extraterrestrial life meanwhile was intensifying, not just on Mars. The moons of Jupiter and Saturn, as well as the upper atmosphere of Venus, were all in consideration. Neighboring star systems also showed promise. But to reach the latter, chemical propulsion simply would not suffice. Those systems were far too weak and inefficient. Even at their fastest, it would take them tens of thousands of years. But there were other options in very early stages of development. Among them sailing.

Humanity has, in fact, long looked to the wind for its voyages. Polynesian explorers, for example, were remarkable navigators who settled distant islands, including Hawaii and New Zealand, while scattering across the Pacific. Perhaps they could offer modern space explorers a few tips. Only this time, instead of the wind, spacecraft would harness the power of the Sun. It was a concept that had its origins as far back as the early seventeenth century, when the earliest hints could be traced to the writings of Johannes Kepler.

At that time, the German astronomer and mathematician had written to his Florentine counterpart Galileo Galilei, who was challenging the Catholic Church's long-held doctrine of an Earth-centered universe through his own telescopic observations. In that letter, Kepler remarked about a comet that had just traversed Europe's skies. Watching it streak overhead, he noticed the comet's tail pointing away from the Sun, which made him wonder whether the Sun's light was exerting a force, like the wind against a sail. Could sunlight then be employed like wind to traverse the cosmos in much the same way sailors had navigated the open seas?

"Provide ships or sails adapted to the heavenly breezes," Kepler wrote, "and there will be some who will brave even that void."

Over the years, the idea would linger as pure theory before, two and a half centuries later, the concept again gained ground with the Scottish physicist James Clerk Maxwell. It was Maxwell who determined that light itself was actually an electromagnetic wave that had momentum in the form of energy-packaged photons, which did, in fact, exhibit "a push." Taken a step further, if a thin reflective sheet could be employed in the near vacuum of space, where a lack of friction allows objects to gain speed without losing much momentum, the notion of a "light sail" could indeed edge closer to reality. By the early twentieth century, the Russian rocket pioneer Friedrich Tsander actually proposed the idea for space travel, raising it in Moscow in 1924 at an exclusive club he had helped found called the Society for Studies of Interplanetary Travel. In those early days of rocketeering, the forum served as a platform for scientists to discuss possibilities of future cosmic journeys, and included luminaries like the Russian research scientist in aeronautics and astronautics Konstantin Tsiolkovsky, a man widely considered one of the founding fathers of modern rocketry. It was Tsiolkovsky who determined Earth's escape velocity and devised initial plans for a multistage rocket. And it was also Tsiolkovsky who said, "Earth is the cradle of humanity, but one cannot live in a cradle forever."

Still, it would be several decades before practical technologies would catch up with their ideas. In 2005, Cosmos 1, a $4 million vehicle put forth by Cosmos Studios and the Planetary Society, sought to test the first solar sail in space, but was destroyed after its Volna booster rocket crashed back to Earth a mere eighty-three seconds after launch. Five years later, it would be Japan's IKAROS that would etch its name into the history books, traveling past Venus in December 2010, roughly half a year after its initial deployment. Then, less than a decade later, a SpaceX Falcon Heavy rocket would launch into orbit a crowdfunded spacecraft known as the Planetary Society's LightSail 2, which became the first space vehicle to demonstrate controlled steering to adjust its orbit using the Sun's photons, employing an onboard motor to unwind some 345 square feet of Mylar sail, roughly the size of a boxing ring and less than the thickness of a human hair, before completing eighteen thousand orbits around Earth. By 2022, the technology had sufficiently advanced that NASA's Near-Earth Asteroid Scout (NEA Scout) used a solar sail to embark on a mission to visit a near-Earth asteroid.

Meanwhile, innovations in materials science and the ability to break up the sail into component parts, which allowed each portion to be articulated in relation to the Sun, afforded a more manageable craft that maximized acceleration without fuel. As it had been for those early Polynesians, the age of solar sailing was now under way.

And yet for Pete, that wasn't enough. He had bigger plans, as did NASA.

· CHAPTER 39 ·

# INTERSTELLAR VOYAGERS

Solar sailing depends on light. And therein also lies the problem. While the push of photons can cumulatively translate into massive increases in velocity over time, the farther the sail is from the Sun, the less power it harnesses. Granted, in the near vacuum of space, the concept of friction is very different from that on Earth. Once a spacecraft is up to speed, it will generally continue at those speeds for the foreseeable future. Still, if humanity wanted to solar sail itself to another star system, the craft would at the very least need tremendous velocity in those early stages. And the craft itself would have to be exceptionally small.

So, in 2016, an Israeli, Russian-born billionaire and science philanthropist named Yuri Milner sought to lay the groundwork for precisely that. On April 12, Milner assumed the speaker's lectern at a press conference at One World Observatory in New York City. To his left sat Stephen Hawking; the producer of the documentary *Cosmos* Ann Druyan; chair of Harvard's astronomy department, Avi Loeb; former astronaut Mae Jemison; physicist Freeman Dyson; and Pete. In 1960, it was Dyson who suggested the possibility of distant megastructures that enclose a star and harness its energy, later bearing his name becoming known as Dyson spheres. Searching for those kinds of technological signatures, therefore, could theoretically be one way of detecting an advanced alien civilization, he suggested. The point of today's press conference, however, was not that. Instead, it was meant to unveil an immensely ambitious plan to bridge a more than four-light-year divide between star systems. The team intended to use swarms of tiny, robotic craft.

"Space travel as we know it is slow," Yuri began. "But, ladies and gentlemen, there is a technology just over the horizon which can get us to the speed we need. . . . For interstellar travel on human timescales, we need a much stronger wind and a significantly smaller probe."

Milner then went on to describe the evolution of three trends in tech that could make a trip to Proxima b more plausible: (1) microfabrication, or the building of miniaturized structures at micrometer scales; (2) nanotechnology, or the designing of such structures by manipulating atoms and molecules; and (3) photonics, or the study of how light emits, transmits, modulates, and can be amplified, processed, and detected.

When it comes to nanoscale production, as we adapt for space travel with smaller and more efficient systems, using infinitesimally smaller probes to scour the cosmos at mind-bending speeds, perhaps the nature of those adaptations will also change the ways we look for life, he seemed to suggest. Astrophysicist Zaza Osmanov indeed pondered whether the Fermi paradox, which points to both a vast universe and a lack of other intelligent life, was at least in part a result of our own inability to look close enough. Perhaps nanoscale craft and indeed civilizations already exist in the near universe, and we are just unable to see them, given energy efficiencies tend to suggest smaller and smaller structures and vehicles. As our own technology advances, so too may our methods of searching require an update.

Meanwhile, Milner was by now holding something between his fingers. It was a processor and a spacecraft—something he called StarChip. Despite its size, he explained how this gram-scale craft, the size of a wafer, could house cameras, photon thrusters, and a power supply, as well as navigation and communications equipment. The device, frankly, looked like an oversized postage stamp. But conceivably, it could one day, become a "fully functional space probe that can be held with two fingers and can be produced at the cost of an iPhone."

"The one I'm holding is a first version with limited functionality, and there will be many iterations to come," he explained, having put $100 million toward a research program to effectively support what he described as an "interstellar sailboat."

It was the most any individual had ever put toward achieving interstellar travel. But in the end, it wasn't quite enough. The project appeared to have stalled, highlighting at least two factors: (1) the difficulty of the task, and (2) the limits of entrepreneurial endeavors in space that showed no signs of a return on investment. This is where America's commercially driven approach to space might fall short, whereas the deep-pocketed and patient capital of state-run programs, like those in China, might be better positioned. Just to make it work required not just the building of nanocrafts and lightsails, but far more expensive Earth-based infrastructure and technologies.

Light from the Sun, despite being the most powerful object in the solar system, wouldn't be enough. Achieving the fractions of light speed needed to accomplish interstellar travel in mere decades would require a more focused and directed power source. To solve for that, the nanocraft would need to get a laser boost, using an array of ground-based lasers that would collectively fire at the sail, pushing it to roughly one-fifth the speed of light—or roughly 134 million miles per hour—in just those initial minutes, and keep firing with as much as a 100-gigawatt-level beam until it was effectively out of range. Theoretically, the craft would then be expected to arrive at Proxima Centauri in approximately twenty years and capture images and other scientific data that it could then transmit back to Earth through beams of light. The idea was to have multiple redundancies and to send "hundreds, maybe thousands of them, to one target" in the event some failed or were destroyed along the way.

Through a swarm of robotic probes, a "mesh network" could also potentially envelop the star system and "generate an optical signal strong enough to cross the immense distance back to Earth," or perhaps serve as humanity's first probes to fan out across the galaxy in search of habitable conditions or indeed signs of life. Perhaps with the advent of 3D printing, AI, and ever more autonomous systems, such probes could eventually become self-replicating, exponentially expanding our human/machine reach and subsequent mappings and communications across the cosmos. These, however, were hugely expensive ideas, which ultimately required a degree of government buy-in at a time when American policy makers seemed happier relying on a private sector approach.

Obviously, even just the idea also wasn't without its technical challenges, including the effectiveness of ground-based laser beams, cooling and navigational mechanisms, questions of sail integrity from impacts with interstellar dust and debris, power generation and storage on board, and how the craft could operate once it reached its destination, let alone broader questions of what it meant to devise autonomous systems with so wide a mandate.

And yet merely reaching distant star systems was not the end goal.

In fact, when I followed up with Pete Worden, it was clear that getting to Alpha Centauri was just an initial step. He also seemed confident that, with the right access, a much bolder plan could be achieved.

· CHAPTER 40 ·

# PARTY IN THE VALLEY

For most of his career, Pete had been a government guy, a military man and civil servant. But he also ran NASA Ames in Silicon Valley. And that, he explained, "tends to get you invited to A-list parties."

There, "I discovered how *not* to approach the billionaires. You learn that these people have lines of people around the block trying to give them ideas and get their money. So it has to be done with more delicacy. It's an art form, and everybody's different." Then at a party in the Valley, Peter found himself surprised to see one of his old team members, Will Marshall of San Francisco-based imaging company Planet, and his co-founder Robbie Schlingler. Will Marshall had earned a PhD in physics from Oxford and worked with the science team that confirmed the presence of large quantities of water on the Moon, then went on to found Planet, which produced high-resolution satellite images of the entirety of Earth's landmass.

"How did you guys get here?" Pete asked.

"Yuri [Milner] invited us."

" 'How do you know him?' Because I didn't know him. And they said, 'Well, he's one of our first investors.' And they said, 'Do you want to meet him?' I said, 'Yeah, I certainly do.'

"And so they took me over, and of course he's got this soft Russian accent. And he said, 'Pete Worden, I've been meaning to get in touch with you.' And so within a few months, I'd met with him a few times and he said he had an interest, not just of rewarding science," as Yuri and his wife, Julia, had done with the Breakthrough Prizes, considered among the world's largest scientific awards. "He [also] wanted to do science," Pete explained. "And over the next year or two, I decided that I was willing to come and run that for him." But before he did, he would also meet with those like Musk and Google's then-CEO, Eric Schmidt, a man who years later, in March 2025, would not only take a controlling interest in the California-based rocket company Relativity Space but also take over as its chief executive. America's billionaires Ghaffarian, Musk, Bezos, and now Schmidt—three of whom had started in software—were effectively emerging as the drivers of America's race in space against China, at the very least to capitalize on the industrial growth surrounding Earth's orbits, but also to dabble in endeavors well beyond. Even Microsoft's co-founder Bill Gates had led a $65 million funding to back a Washington-based reusable launch company called Stoke Space, in an apparent bid to begin to leverage space through emerging satellite technologies. But for Pete, the question now focused more on how to leverage those contacts into his broader deep-space ambitions, particularly when it came to applications of emerging quantum technologies—which stood to benefit, especially if a stable supply of helium-3 could ever be sustainably extracted from the Moon.

"I had some quantum experts that wanted to do quantum computing. And interestingly enough, they were all either Russian or Ukrainian," he recalled. One of those experts was Vadim Smelyanskiy, who led the physics modeling group in the Computation Science Division at NASA Ames.

"He would come in and he'd say, 'You are a physicist. Let me show you quantum computing.' And he had pages of damn equations. And I said, 'Yeah, I'm an observational astronomer. Just tell me what this does.' Finally, I managed to get him to tell me in English and not in mathematics."

Essentially, Smelyanskiy had been working on something called failure mode development and fault management. "When you have abnormal behavior in a flying rocket," he explained, "the root causes of that and the consequences and the timing, as well as the instrumentation that is needed to predict and diagnose and isolate the causes, are all based on physics." But there were some inherent limitations when it came to conventional methods and classical computing. "With traditional physics disciplines, somebody is doing composite materials or structural mechanics, and somebody is doing electromechanics, or computational fluid dynamics. In a situation of a complex integrated physics environment, however, where mechanical, propulsion, electromagnetic instrumentations are all tied together, there's no such possible software code that would integrate all these multiphysical simulations in one coherent fashion. There are too many interfaces. It's a huge inhomogeneity of physics processes—a variety of special skills, and diversity of those processes. It's like trying to describe a human being through one physics model that couples the mechanics of your motion and what happens in the brain down to the neuron level. That's simply not possible."

And yet quantum computing offered exciting new possibilities, simultaneously exploring multiple solutions. Fields once constrained by linear processing suddenly seemed on the precipice of explosive growth. And for those like Pete, and many others, "true AI would only occur when we have quantum machines," with uses that held both immediate and long-term potential applications. For instance, for the past eight decades, NASA Ames has been heavily involved in air traffic control, innovating solutions and performing related research in a bid to optimize the flow of traffic of commercial airliners.

"You have thousands of airplanes simultaneously. How do you optimize this for the best routes, the maximum fuel efficiency, and maximum safety?" Pete explained. "That you can't fully solve computationally. But a quantum machine is different because you can think of it in that each quantum bit computes in all universes simultaneously as a probability function. So if you add an extra city, it's one extra quantum bit."

Whereas classical computers employ binary bits represented as either 1s or 0s, quantum computers are not similarly constrained, and can operate in what's called a superposition, or both 0 and 1 simultaneously. That allows quantum computers to process information in immense amounts relative to traditional computing. Quantum computers use qubits, or quantum bits, as their basic unit of information. Plus, the phenomenon of entanglement, in which the state of one qubit is linked to another, regardless of the distance between them, could work to dramatically expand computational power. In fact, by February 2025, Microsoft had announced its new Majorana 1 processor had achieved a "transformative leap toward practical quantum computing," which seemed to only further ambitions when it came to things like cryptography, materials science, and pharmaceutical development, to name just a few.

"In the same way that the invention of semiconductors made today's smartphones, computers and electronics possible, topoconductors and the new type of chip they enable offer a path to developing quantum systems that can scale to a million qubits and are capable of tackling the most complex industrial and societal problems," the company said in a statement.

By exploring a multiplicity of solutions simultaneously through ultrafast, ultrasecure quantum networks, the concept was akin to the difference between water emerging from a pipe, bit by bit, and ocean waves' more complex and interconnected nature.

And back then, Pete saw a mutual benefit between what he wanted to achieve at NASA Ames and where those like Google and Microsoft were ultimately headed.

"I think from the Google side, there was commitment and understanding that quantum computing, if they don't lead in that in their business, they'll become obsolete," he explained.

To better understand that evolution, however, we must rewind the clocks more than a decade to the conversation Pete had with Eric Schmidt, who—for transparency's sake—could not be reached to independently confirm the following statements, which are therefore based solely on Pete's own recollection of events.

"Maybe if Google bought one of these and we housed it at Ames, we could form a joint project," Pete went on to explain. "Eric looks through this thing and we talked to him, and he asked a few probing questions. He said, 'Look, I want to do this.' He then added, 'But I really need to have Larry and Sergey's approval,'" referring to Google's co-founders, Larry Page and Sergey Brin.

"So as we walk out from his office, Larry's coming by and he sees me there, and with Eric, he said, 'What are you guys doing?' And so Eric has this set of charts, and so we went over it and Larry said, 'I think that's fine.' And Schmidt says, 'Well, there you have it. I guess we're going to do it.'" The partnership was now set, effectively adding a series of collaborations between the search giant and NASA Ames, perhaps most notably the ambitious bid for quantum supremacy, or the ability to carry out calculations that classical supercomputers cannot.

By 2018, researchers in Silicon Valley seemed confident. In a post on a Google Research page, a Google Quantum AI Lab research scientist named Julian Kelly said that the team was "cautiously optimistic that quantum supremacy can be achieved with Bristlecone." A little more than a year after Kelly's post, Google announced that it had indeed achieved that milestone of "supremacy," albeit with its less powerful 53-qubit Sycamore processor.

"In reaching this milestone, we show that quantum speedup is achievable in a real-world system and is not precluded by any hidden physical laws," the authors wrote in the journal *Nature*.

Q Day, as it's often called, may soon be the day when large-scale quantum computers effectively render encryption systems obsolete. Such a large-scale quantum computer, as Arthur Herman, a senior fellow and director of the Quantum Alliance Initiative at Hudson Institute, notes, might be considered "the ultimate thermonuclear device in cyberwarfare," and is an area Beijing continues to prioritize as a key emerging technology—something with profound implications for securing satellite communications and space-based infrastructure.

But, seemingly as always, Chinese scientists were also fast making strides of their own.

"We are improving the precision of quantum computers, while expanding their applications such as in encryption techniques and precious measurement," Yu Dapeng, of the Chinese Academy of Sciences, explained in October 2024. By 2024, Beijing's heavy investments in quantum technologies had paid off with the development of a 504-qubit computing chip, called Xiaohong. In May of that same year, researchers at the University of Science and Technology of China had announced the development of a new "Plasmonium" qubit, with sweeping potential applications in data processing, compact hardware, and algorithm optimizations useful in enhancing the kinds of AI-driven systems that probes to other star systems would likely require. That application would also represent a significant security threat in an escalating quantum race that was quickly spilling into its broader space competition with the United States.

"If these technologies are built in China, for example, they are not necessarily going to follow our privacy rules or our ethics," Schmidt would later tell Fareed Zakaria during a 2021 interview. "We have to be careful to win this battle."

Pushing the concept forward as it related to space exploration and discovery, however, was where Pete's real interests seem to lie. If the goal, in other words, was to send intelligent machines to scour other planets for possible life, "it'd probably have to use quantum computing." And yet what Pete seemed to envision wasn't just a leap of tech. Rather, it was a radical redefinition of our species' own relationship with the universe, in which autonomous, quantum-powered machines would become our eyes, ears, and minds in the deeper chasms of space. The technology itself, however, was also poised for an explosion in capacity. Picture, for instance, the prospect of quantum communication channels reaching across interstellar distances, or telescopes equipped with quantum optics, and their ability to enhance our understanding of how photons interact with individual atoms, exponentially improving the ability to examine the very origins of the universe. Then picture quantum leaps in materials sciences and computational chemistry, running highly complex simulations in a bid to render ever more optimized space vehicles, while also mapping the solar system and interstellar systems with exponentially more detailed imagings of stars, exoplanets, and asteroids.

Pete sensed Eric was interested. He had laid the groundwork early and was ready to use his agency cred to move the needle. But still, Schmidt needed to be approached carefully. At one point, somewhere around 2008, Pete saw his chance. Google's CEO mentioned that he had "never seen a shuttle launch."

"I'm sure we can arrange it," Pete replied.

"Not very many people get fifteen hours of uninterrupted time with the CEO of a company like that," he later told me. The two conversed. And Pete explained that if you "want to send a robot . . . to look for life, I'd really like to send an intelligent robot [that would] probably have to use quantum computing."

Given the yawning round-trip communications lag, the capacity of the machines to operate autonomously would have to take primacy. Over the years, those conversations would also involve Yuri Milner and Saul Perlmutter, who in 2011 had become a Nobel laureate, along with the astronomers Brian Schmidt and Adam Riess, for their work in surmising how dark energy served as a repulsive force that contributed to the expansion of the universe. Meanwhile, with growing advances in launch technology, specifically Starship and its potential to carry as much as 150 metric tons into orbit in reusable mode, new possibilities were rapidly coming online.

"Starship enables us to launch . . . ground-based telescopes into space," explained Pete, which could better avoid atmospheric distortion for scientific and exploratory purposes, but also potentially be useful in communications across such vast distances. By launching three, astronomers could also dramatically improve mapping and image resolutions gleaned from interstellar space through triangulation and something called interferometry—which is basically just the combining of signals to gain more detailed measurements of distant stars and planets. Such "terrestrial light buckets" could then theoretically better catch optical signals sent from distant star systems. AI-powered probes suddenly seemed like just the beginning. Combined with advances in bioengineering, especially synthetic biology, which involves creating new biological systems not found in nature, what once seemed like science fiction now appeared to be a growing possibility.

"Sometime in the next century or two, we could probably put a payload on a planet orbiting a nearby star, and that payload could boot up life," Pete said matter-of-factly. Taken a step further, it's then reasonable to speculate that if humans could achieve this, it could also be reflective of humanity's own origin story—an iteration of the panspermia theory that suggests life on Earth might have originated elsewhere, with organic molecules having hitched a ride aboard a comet or meteorite traveling through the cosmos and ultimately striking Earth. If, however, we now hold the nascent abilities to initiate life artificially, would it then be possible that others did the same for us, effectively kick-starting our own beginnings from afar. It seemed both exciting and alarming. Just the prospect of bringing human-devised life to another star system as an initial step, perhaps as some sort of biological redundancy, seemed a testament to the converging of rapidly advancing disciplines and pioneering spirits. But those should-we-actually-do-this questions were simultaneously growing.

What kind of synthetic biologic might be delivered? In what ways could AI-fueled, quantum-powered autonomous systems develop? Was that even knowable? Plus, absent the kinds of entanglement communications that might bridge that distance, with an eight-year communications lag between star systems, how could humans ever hope to keep tabs in a timely manner? How would we know if Asimov's law, which sought to create an ethical structure of guidelines that governed AI-powered systems, were ever violated?

Pete, however, wasn't done talking.

"It would probably have a quantum brain," he added. In fact, a growing body of research seemed to suggest, albeit controversially, that cognitive human brain function and our own notions of consciousness could indeed rely—at least in part—on quantum processes. The idea is hotly debated, with some notions, like orchestrated objective reduction proposed by Roger Penrose and Stuart Hameroff, arguing that quantum effects within neurons may play a role in consciousness. Even Erwin Schrödinger, the Nobel Prize–winning physicist who contributed mightily to the development of quantum mechanics, said that "consciousness cannot be accounted for in physical terms." From there, philosophical considerations would give rise to a range of ideas about the creation and evolution of life itself, especially amid quantum-fueled advances in fields like bio-additive manufacturing, with labs already printing functional tissue structures for research and medical applications.

"It would boot up life with printout genomes," Pete added, a reference to advanced DNA synthesizers that could theoretically employ 3D bioprinters to assemble genomes in cellular structures on a new host planet, after being presumably cryopreserved over the course of their long journey from Earth.

"That's my ultimate objective," he added. "So when you ask me am I interested in making money in low Earth orbit? No, I don't consider that even space.

"My interest is the nearby stars."

That idea, of course, was fine for seedlings and nanocraft. But what about human-rated ships? Was that even possible? Assuming humans even had a habitable place to go to, the mass of a craft would still pose major propulsion problems. Given it would take more than six millennia to reach Proxima Centauri by current chemical propulsion methods, the mission would then demand large crews that procreate over a multigenerational voyage. The ideas were many. Could asteroids theoretically be converted into vessels, using additive manufacturing and in situ resource utilization, to satisfy size requirements that might otherwise be too large to launch from Earth? In fact, NASA commissioned a study in 2016 to review feasibility. The craft itself would need things like rotating habitats to generate artificial gravity and enhanced shielding against high-energy cosmic rays to protect the crew from radiation. Still, the idea of a thousands-year journey that depends on scores of generations of uninterrupted human procreation inside a ship transiting the vastness of space might not be the most reliable way to settle distant star systems. There are, to some extent, analogies for such undertakings. Consider those generations of cathedral builders who never saw the completion of their efforts. Notre-Dame de Paris, for instance, began construction in 1163 and wasn't completed until 1345, meaning there were generations of workers who would spend their entire lives in toil and never see its completion. But these were ships. And so chances were—to make it really work—they would have to travel a lot faster than anything humanity had ever created.

· CHAPTER 41 ·

# NUCLEAR ROCKETS

When reporters revisit President John F. Kennedy's address to a joint session of Congress in the spring of 1961, they often frame it as a bold moment of promises made—and ultimately kept. The president vowed to land a man on the Moon before the decade was out, and return him safely to Earth. And while Kennedy himself would not live to see that vision fulfilled, his challenge nevertheless ignited an American space effort that stretched the very boundaries of human exploration.

Curiously, however, the Moon wasn't the only goal or commitment outlined. "An additional 23 million dollars, together with 7 million dollars already available, will accelerate development of the Rover nuclear rocket," Kennedy added just moments after his call to place astronauts on the lunar surface. "This gives promise of someday providing a means for even more exciting and ambitious exploration of space, perhaps beyond the moon, perhaps to the very end of the solar system itself."

Yet by the time his words echoed through the legislature, NASA and U.S. defense agencies were already exploring the use of atomic energy to accelerate spacecraft at speeds that chemical propulsion couldn't match. Almost simultaneously, those same agencies were developing nuclear-thermal engines through initiatives known as Project Rover and Project NERVA.

By the mid-twentieth century, ideas about the utility of nuclear weapons began to evolve. A Polish-born mathematician who had worked on the Manhattan Project was now exploring novel uses of nuclear explosions and how they might be applied in the vacuum of space. For Stanisław Ulam, the concept—rooted in fluid dynamics, shock waves, and the sheer power and mechanics of nuclear explosions—was essentially based on the controlled detonations of a series of hydrogen bombs behind a spacecraft, which could then theoretically ride the resulting push of high-energy photons to achieve fractions of light speed. (You would be forgiven if your mind suddenly conjured images of Slim Pickens waving his cowboy hat atop a nuclear bomb in Stanley Kubrick's iconic 1964 film, *Dr. Strangelove.*) Working alongside the Hungarian American theoretical physicist and chemical engineer Edward Teller, he developed what is known as the Teller–Ulam design, a two-stage model credited with the large-scale development of hydrogen bombs. But as for space travel, the model would be decidedly different from traditional rocketry.

Rather than rely on continuous rocket propulsion, the ship could in theory achieve ever-increasing speeds as hydrogen bombs were detonated in a sequence behind the craft. The concept, in fact, showed up in a 1955 Defense Department report, titled "On a Method of Propulsion of Projectiles by Means of External Nuclear Explosions." Authored by Ulam and C. J. Everett, it noted that "repeated hydrogen bomb explosions outside the body of a projectile are considered as providing means to accelerate such objects." A "pusher plate," affixed behind the craft, would supposedly absorb the shock and propel the craft forward.

"The critical question about such a method concerns its ability to draw on the real reserves of nuclear power liberated at bomb temperatures without smashing or melting the vehicle," the report noted. To help address this, solid disks would be ejected in between the bomb(s) and the ship. Again, this was a far different time, replete with the kind of risk-taking that is almost anathema to our more modern age. But the explosion would then theoretically render those disks into a plasma that pushed against a ten-meter saucer-shaped plate attached to the rear of the craft. The design was meant to effectively catch "all or most of the exploding propellant." And the resulting momentum shift would, again theoretically, send the craft careening out into space at almost unfathomable speeds. That is, of course, if the ship wasn't vaporized in the process, if the bomb's resulting electromagnetic pulse didn't fry the craft's circuitry, and if the massive spew of radiation didn't kill everyone on board. Like I said, it was a much different time.

Still, as the team expanded and included those like Ted Taylor, a theoretical physicist at General Atomics, a division of General Dynamics founded in 1955 tasked with exploring peaceful uses of atomic energy, Project Orion was created, which sought to formalize both NASA's and the U.S. military's explorations into exploding nuclear bombs as a means of spacecraft propellant. Almost simultaneously, Project Rover and Project NERVA were also devised to explore nuclear-powered missions to Mars, as well as potential upper-stage uses for the Apollo program. While Orion generally oriented more toward propulsion systems powered by external detonations, Rover and NERVA centered more on nuclear-powered engines. But momentum is a curious thing, especially when it comes to American space policy. And within just a few years, it seemed clear that much of the space community was now beginning to move away from those more daring concepts. By 1958, the United States had already begun studying the high-altitude effects of nuclear explosions, including the aforementioned Starfish Prime. Yet the concept of nuclear-powered spacecraft never quite disappeared. And for good reason.

The potential of nuclear power as a compact and long-lasting energy solution is difficult to ignore, given the value of low-mass, high-energy fuel sources in space. Unlike solar panels, which tend to struggle during lunar nights or in the dim outer reaches of the solar system, nuclear has the capacity for longevity at a distance.

It's important, however, to put these more modern nuclear efforts into some context, given it was Kennedy who generations ago called for "a major national commitment," which included nuclear-powered missions. Perhaps it was merely born of a burgeoning Cold War era with the Soviet Union, in which U.S. policy makers had the funding wherewithal for such big ideas and unproven initiatives. At its peak, for instance, the Apollo program would consume nearly

4 percent of the national budget, and would come on the heels of today's equivalent of a $30 billion Manhattan Project that ushered the world into the atomic age. And yet, in the intervening decades, such bold promises for harnessing nuclear power in space would fall far short of those early ambitions.

"We've spent nearly $20 billion on nuclear [space] power since the '50s, and the only system we currently have is a lightbulb-sized, 100-watt radioisotope generator," Bhavya Lal, NASA's former associate administrator for technology, policy, and strategy, told me one afternoon in 2025.

Voyager 1 and Voyager 2—the only two spacecraft to have reached beyond our solar system—employ the natural decay of plutonium-238 in their radioisotope thermoelectric generators to convert heat into electricity. NASA's New Horizons interplanetary space probe, designed to study Pluto and its moons, also employs power through the natural radioactive decay of plutonium dioxide fuel, comparable to that of a household lightbulb in terms of wattage, though hardly the sort of power that could lessen travel times to the Red Planet.

"Beyond Mars, we really don't have a whole lot of solar power," and "there is no way to do a whole lot of science without nuclear," Bhavya added. "We went to Pluto with a New Horizons mission [and] the probes whisked past Pluto. Why? Because we didn't have the Delta V to go into orbit. So . . . all we have is basically twenty-four hours' worth of data, which took a year to get back. [Because] we had a lightbulb amount of power, and we had to split the power between sending data back and doing more science or processing data."

Meanwhile, as NASA's own landscape seemed to be shifting, in part away from federally funded science, commercial entities were exploring how their efforts might align with emerging policy priorities, and where those nuclear efforts might fit in. In April 2025, Kristin Houston, president of L3Harris's space propulsion and power systems, told the *SpaceNews* journalist Sandra Erwin that "we are finally at the cusp for both nuclear electric propulsion and nuclear thermal propulsion," and that the company was monitoring

NASA's Fission Surface Power program, meant to develop nuclear power systems for both lunar and Mars surface operations. Fission propulsion also suggested a potential shortening of the months-long journey it took to reach the Red Planet—a phenomenon both Beijing and Moscow were now also actively exploring. Nearly a dozen research institutions and universities across China were exploring the prospect of scaling down a megawatt nuclear reactor to fit inside a Starship-like vehicle, just as Russia's main space agency, Roscosmos, detailed plans for its own craft to be powered by a similar reactor. The two nations had also discussed collaborations on a lunar reactor, which could be launched as early as 2033, to power the joint lunar station that by then was expected to be coming online. Nuclear energy in space was clearly emerging as a key future technology in this now burgeoning space race. But as the tide of technology advanced, and conversations of what was possible beyond our solar system advanced with it, even fission's fire seemed ill-suited. Perhaps an even more powerful kind of reactor was needed to shorten the trip—something that could harness the very power of a star.

Fusion (a still-in-development technology) could theoretically pick up the pace. In 1968, British theoretical physicist Freeman Dyson had even suggested using fusion energy pulses to travel between star systems. But the energy had never been practically harnessed for peaceful purposes, mostly because of the difficulty in sustaining fusion long enough to produce more energy than it took to initiate the process. The Sun effectively accomplishes this by the use of gravity, which keeps its structure and sustains fusion reactions at the core. On Earth, research on controlled fusion reactors often centers principally on the use of inertia and/or magnetic fields. But it was the machines' sustainability that was key, even if questions of practical fusion propulsion systems at the very least seemed decades away. Then again, with the kinds of development being wrought by AI and the sorts of quantum systems that companies like Microsoft were pioneering, the pace of innovation and technological breakthrough were now poised for a supercharge. Taken together, these were no longer isolated projects. Instead, they were threads in the tapestry of frontier science that promised to offer almost unrivaled power to the nation or groups that could corral them.

Sustained fusion energy, meanwhile, was already being worked on by a host of nations. In fact, a sprawling facility nestled in the South of France had by 2007 already broken ground. Known as the International Thermonuclear Experimental Reactor, or ITER, it had become the world's largest fusion reactor, and among the most expensive and complex machines ever built, a beast of a structure that towers above the French countryside like some sort of metallic cathedral to the future.

France, therefore, seemed like the next logical stop on this journey.

· CHAPTER 42 ·

# HARNESSING THE POWER OF THE SUN

It was Mikhail Gorbachev's idea. Or, perhaps more likely, it was a cadre of Soviet scientists who convinced the Soviet premier to propose it with U.S. president Ronald Reagan. Yet it was France that would emerge as the natural home for this ambitious and hulking new reactor, effectively designed to harness Promethean fire. After all, the French have long loved experimenting with innovative forms of power. That penchant dates back to at least the late 1940s, when the country launched its postwar nuclear program, one that President Charles de Gaulle would later champion. Today, nuclear power generates roughly two-thirds of the nation's electricity. Yet the engineers and scientists at ITER were up to something considerably bigger, the product of a joint effort by the European Union, Japan, India, the United States, Russia, China, and South Korea that sought to devise a form of fusion energy—essentially the makings of a kind of artificial star on Earth.

Situated in Cadarache, a forested enclave in the Provence region of southern France, ITER is thought to be Europe's largest hub for energy research and innovation. The sprawling facility is home to a thirty-meter-tall tokamak, which represents the core of the reactor, in a bid to create a virtually limitless supply of clean fusion energy.

Following a November 1985 Gorbachev meeting with President Reagan in Geneva, the subsequent communiqué emphasized "the potential importance of the work aimed at utilizing controlled thermonuclear fusion for peaceful purposes." And, it added, much like the words etched upon the lunar plaque of the Apollo 11 mission,

Part of a tokamak fusion reactor being built in France as an engineering megaproject aimed at creating energy through a fusion process similar to that of the Sun, 2022.

its use would be for the benefit "of all mankind." Fusion power, if it could be made operational, held immense promise. It offered an essentially inexhaustible energy source, and employed more widely available hydrogen isotypes, such as tritium and deuterium, the latter readily extractable from something as common as seawater. Fusion reactors were also thought capable of producing four times as much energy per kilogram as their fission counterparts, and were also generally safer. Unlike fission, a fusion reactor's malfunction does not risk the same sort of runaway chain reaction or meltdown, which coincidentally occurred in Chernobyl just one year after that Geneva communiqué. Despite all that promise, however, the pace of development in France would remain maddeningly slow.

Decades earlier, in the United States, Ulam and Teller's contributions to the creation of a thermonuclear fusion weapon in 1952 had already showcased the power of fusion, with isotopes of hydrogen fused to create helium under immense temperatures and pressure, a process triggered by the detonation of a conventional fission bomb. But the question of whether those reactions could be produced in a more sustained fashion to generate continuous power seemed to be the real question of ITER's future. The science behind this mimicker of nature had to be overcome.

For example, within the Sun itself, much of fusion's sustenance comes down to gravity. In fact, under the nebula theory, the creation of the solar system was in part due to the influence of gravitational forces, which drew molecules together within the pre-solar nebula, possibly triggered by a nearby supernova. That clumping in turn increased its own gravity, which led to more clumping. As clouds of molecules collapsed, a dense core at the center of the system known as a protostar was forged. Gravitational forces then further compressed the core, and temperatures swelled until fusion occurred. But it is the Sun's own massive size, and the resulting inward gravitational pressure it creates, that effectively balances the fusion energy it generates. As a result, the Sun neither expands uncontrollably nor collapses as it fuses roughly 600 million metric tons of hydrogen every second.

It is that balancing act that scientists on Earth have been trying to replicate. And in 1950, they got a bit closer. The Soviet physicists Andrei Sakharov and Igor Tamm proposed a magnetic confinement system using high-temperature plasma, and continued experiments at the Kurchatov Institute in Moscow. The concept essentially employed magnetic confinement to move atoms as close as possible to increase the possibility of collisions. Once the atoms struck each other with sufficient force, they could conceivably overcome their own electromagnetic barriers, thereby allowing for the creation of fusion energy. Fusing atoms was, of course, also demonstrated with the hydrogen bomb in 1952. But unlike the spike of energy produced from a bomb, those more sustained fusion reactions required a continuous input of energy to heat the surrounding plasma, along with cryogenics for the coils that enable superconductivity. The concept was reportedly classified until 1956, when it was presented at a conference in Harwell, U.K., and later published by Pergamon Press in 1961. And yet one of the first major breakthroughs occurred less than a decade later in the Siberian city of Novosibirsk.

There, nestled near the banks of the Ob River, the first so-called tokamak systems that confined plasma using magnetic fields were tested. It would be one of several that could heat plasmas to more than 100 million degrees Celsius. But the one at ITER was to be the world's largest, roughly twice the size of its nearest competitors, with an ability to house as much as ten times the volume of plasma.

Given ITER was born of fusion research in part developed by Soviet scientists, it seemed fitting that it was a Russian engineer who greeted me upon arrival in Cadarache.

"Here," said Vladimir Tronza, handing me a helmet and gloves.

"They always are strict about them," he said with a smirk. "Even if you don't touch anything."

With his PhD from Moscow State University of Instrument Engineering and Computer Sciences, he was now employed not only as manufacturing and assembly engineer at ITER but also apparently as my tour guide for the day. Standing in front of ITER's twenty-four-hundred-acre complex is a shock to the senses, likely because of the contrast with the region's more bucolic, southern French countryside.

"When I came here in 2010, basically there was nothing here, just

The International Thermonuclear Experimental Reactor, or ITER, in southern France, 2022. An international team of more than two thousand physicists and other scientists work on building the world's largest nuclear fusion plant in an attempt to harness the type of energy generated by the Sun. With Vladimir Tronza (*right*), a manufacturing and assembly engineer originally from Russia.

a platform when they started excavation," he explained. Meaningful construction, in fact, wouldn't begin for another six years. But just as he spoke, Vladimir stopped to greet a passing colleague. "Hello, Chen," he said, gesturing with his hand as the two men crossed paths. Lu Chen worked as an engineer and liaison of sorts to the broader engineering team of which Vladimir was a part. "He has a background in designing what we call feeders," he later explained. "It's the components that will feed the magnetic system."

ITER appeared to be truly a global effort.

In fact, China supplied each of the site's thirty-one magnet feeders, which worked to relay power, cryogenic fluids, and instrumentation cables to the superconducting magnets that have the unique characteristic of "crossing the warm/cold barrier of the machine." To attain superconductivity, the magnets had to be cooled using cryogenic fluids. The result was extreme colds adjacent to plasmas that burned hotter than the Sun's core, roughly 150 million degrees Celsius, inside the massive donut-shaped machine that looked like something George Lucas might have conjured.

When it came to fusion technology, Chinese scientists were already making remarkable strides. By 2025, a tokamak facility in central China had set records for generating a sustained plasma six times hotter than the surface of the Sun. But this device was meant to also allow scientists to study helium nuclei, produced by fusion reactions. A year earlier, ITER's director general, Pietro Barabaschi, would head off on a diplomatic trip to China, where a Shanghai-based reactor had been built by a Chinese energy company called Energy Singularity. It claimed to be focusing on "high-field, high-confinement and compact tokamak with HTS magnets," and was thought to have given China a first-mover advantage with its completion of a system that could produce the high-temperature superconducting magnetic confinement needed for nuclear fusion. In a bid to make it commercially viable, the company's comparatively low-cost tokamak had been built in an astoundingly fast two years.

At roughly the same time, across the Pacific, researchers at the National Ignition Facility in Livermore, California, had achieved a landmark breakthrough in a different approach to fusion. It was about two years earlier when, in what was hailed as the first controlled fusion experiment to reach ignition, they produced more energy from the reaction than was used to initiate it, a milestone that had been decades in the making, effectively validating research that had long claimed fusion energy could indeed become viable. It also seemed to lend new credence to those who wanted to push the technology further, suggesting fusion energy could even one day power off-Earth habitats and perhaps even propulsion systems. Such rockets could drastically shorten the journey to Saturn to mere months, whereas NASA's Dragonfly mission, for context, is slated to take about seven years to reach Saturn's moon Titan. Such a system could also, again in theory, reduce the voyage to Proxima Centauri from thousands of years to just over a century.

Meanwhile, as ITER seemed increasingly weighed down by delays, a private fusion market was experiencing a huge surge in investment and activity as private companies and public-private partnerships sought to advance the technology. And at least five companies surveyed were making a push toward fusion-related space propulsion as a primary market. Another seven listed space as a potential spin-off market. Companies like Helicity Space, Princeton Satellite Systems, and the U.K.-based Pulsar Fusion were making headlines. And in 2023, the latter two began collaborating on construction of an eight-meter fusion reaction chamber with the intention of conducting both plasma tests and test fires in the hopes of reaching exhaust speeds in excess of 500,000 miles per hour.

"It's irresistible to the human evolution of space," Pulsar's CEO, Richard Dinan, explained. "This is happening, because the application is irresistible." Perhaps the most successful effort had also occurred in the U.K., where the Joint European Torus in Oxfordshire used just 0.2 milligrams of fuel to set a world record for energy output, generating 69 megajoules for 5.2 seconds. For a fleeting moment, that system reached 150 million degrees Celsius, thus briefly becoming the hottest place in the solar system. Efforts to shrink the size requirements were, meanwhile, happening in both the private and public sectors, typically using a process called lattice confinement. This could reduce the hulking size requirements of facilities like those at ITER, which increasingly seemed unable to keep pace with modern commercial efforts, with its $6.3 billion budget by now having ballooned to more than $22 billion.

Unlike tokamak reactors, which rely on intense magnetic fields and extremely high temperatures to confine plasma, lattice confinement fusion involves trapping hydrogen isotopes within the crystal lattice of a metal, which allows those larger metal atoms to serve as a kind of cage that confines the fuel. Researchers then use methods like gamma radiation or ion beams to initiate fusion reactions in these confined spaces. That was a step toward the kinds of size restrictions necessary in space vehicles.

"We always try to remind people that this project was born as an experiment," Vladimir told me as we entered the facility. He was defending it, in a way. "We have all these facilities around the world where we have fusion machines running, but this one will be really a big one."

That, of course, was unmistakably true.

Causeways lined the behemoth skeleton of the machine where we walked. And every so often our path turned in toward a labyrinthine set of white tunnels, which meandered around its colossal chambers. As we chatted, he explained the site's various functions, even if its future remained unclear. The sustainable fire of the stars now, of course, seemed within reach. And yet even *that* could still fall short of the extraordinary demands required to push beyond the boundaries of our solar neighborhood in search of distant, habitable worlds. To cross such vast interstellar distances might require a more exotic mechanism or fuel source, perhaps using one of the rarest substances in the known universe—one that however impractical, held the potential to release three hundred times as much energy per unit of mass as fusion, and more than ten billion times as much as chemical combustion.

Just a few hours' train ride north would bring me to where that substance was being made, in a factory on the outskirts of Geneva.

· CHAPTER 43 ·

# ASK A FISH ABOUT WATER—A VISIT TO CERN

It would have been enough to visit CERN. The world's largest particle physics laboratory is home to the most powerful high-energy accelerator, colliding subatomic particles at nearly the speed of light around a seventeen-mile tunnel. Scientists at the facility, which spans Switzerland and France, had already earned widespread acclaim in 2012 for the discovery of the Higgs boson, a quantum field that fills the universe and gives physical objects their mass. First theorized by British physicist Peter Higgs in 1964, it was billed as almost a sort of cosmic soup that drags on particles, which physicist Brian Greene aptly captured in an article for *Smithsonian Magazine:*

> Physicists tell a parable about fish investigating the laws of physics but so habituated to their watery world they fail to consider its influence. The fish struggle mightily to explain the gentle swaying of plants as well as their own locomotion. The laws they ultimately find are complex and unwieldy. Then, one brilliant fish has a breakthrough. Maybe the complexity reflects simple fundamental laws acting themselves out in a complex environment—one that's filled with a viscous, incompressible and pervasive fluid: the ocean. At first, the insightful fish is ignored, even ridiculed. But slowly, the others, too, realize that their environment, its familiarity notwithstanding, has a significant impact on everything they observe.

Finding it, however, was the tricky part. After protons were accelerated to nearly the speed of light and collided, those high-energy collisions produced a cascade of particles that existed for mere fractions of seconds, before they quickly decayed or broke down into other particles. Confirming its existence was therefore a fundamental step in better understanding the very fabric of our universe. And yet, interestingly enough, it might have been found much earlier—not in Geneva, but in Texas—had politics not interfered with plans for the Superconducting Super Collider in the state's northern town of Waxahachie. With a ring circumference of roughly fifty-four miles, it would have dwarfed the size and power of CERN, had it been completed, effectively cementing American supremacy in the world of high-energy physics.

Yet it was not to be. Shifting political winds and budgetary concerns led to its cancellation in 1993, even though roughly one-fifth of the project had already been completed. Once again, it seemed America's scientific retrenchment had spelled opportunity elsewhere.

Perhaps by now expectedly, it was China that had plans for an even larger accelerator, boasting sixty-two miles of tunnel. The Circular Electron Positron Collider was set for construction for as early as 2027. Once completed, it would constitute the world's largest and most powerful collider, allowing researchers to probe the Standard Model of particle physics in unprecedented ways, including age-old questions surrounding both dark matter and antimatter. It was an ambitious project and a signal of the seriousness of Beijing's efforts. China's own technological prowess drew concerns of a "reverse brain drain" of top-tier scientists who then might be drawn to work on cutting-edge projects of unrivaled scale and capability. In fact, given China already maintained a smaller collider, known as the Beijing Electron-Positron Collider, it wasn't clear if that critical mass in particle physicists hadn't already started to shift in its favor.

Meanwhile, Chinese scientists were increasingly returning home. In 2023, Stanford University researchers analyzing migration trends of U.S.-based Chinese scientists, in fact, found that although China had long been an important foreign supplier of U.S.-based scientists, their departures had increased by 75 percent since 2018, with approximately two-thirds of such scientists moving back to China. The effect—the researchers determined—was in large part tied to the so-called China Initiative, a program developed under the first Trump administration to thwart Chinese theft of American intellectual property. While only a handful of the China Initiative cases resulted in convictions or guilty pleas, the risk was that "U.S. science [would] likely suffer, given the loss of scientific talent to China and other countries."

The issue again surfaced with Trump's second term, this time with stricter scrutiny of America's H-1B visa program, a system that allows U.S. companies to hire skilled foreign workers, particularly in the fields of science, technology, engineering, and mathematics. And by the spring of 2025, American tech companies appeared to be quietly advising their foreign-born employees not to leave the United States, lest they risk being denied reentry; just as calls to end citizenship of those born in the United States fueled mounting anxieties that their future children could become stateless. Those actions, coupled with cuts to university research efforts and broader slashes to federal science programs, including the proposed halving of NASA's science division, represented a "total attack on all of science in America," as Eric Schmidt described at an AI+ Biotechnology Summit in 2025. "We're up against China that is pouring a trillion dollars into this, and we're screwing around with funding the core people to invent our future." Earlier that same year, Meta's chief AI scientist, Yann LeCun, accused the Trump administration of starting a "witch hunt in academia," reminiscent of the old "science in the haunted 1950s" era.

Only this time, with moves that threatened to put a chilling effect on America's talent pipeline, China was in a position to capitalize. Plus, such immigrant policies also seemed to be married with an erosion of global credibility among U.S. partners, exemplified by a remilitarizing Europe and sweeping tariffs aimed not just at adversaries like China but also at the economies of long-standing allies like Canada and the European Union. For the space industry, the makings of a global trade war promised to put new pressures on production costs, adding fresh measures of geopolitical risk onto their balance sheets. Coupled with Beijing's own export suspension of rare earth minerals and magnets critical to the aerospace industry, America's turn from allies and free trade norms, under the goal of compelling a resurgence in U.S. manufacturing, had opened yet another window of opportunity on which Chinese leadership sought to capitalize.

Yet that rise was also far from inevitable. The nation had a clear demographics problem, driven in part by the legacy of its old one-child policy, which was now contributing to a shrinking workforce and newfound pressures on its social safety nets.

An unsteady Chinese housing market and steady loss of labor-intensive manufacturing meanwhile exposed the political vulnerabilities of its one-party leadership as the economy that once lifted millions of Chinese from poverty slowed down. Over the generations, a tacit bargain had been struck, in that China's authoritarian-style rule would be tolerated provided an overall quality of life increased, an inherent vulnerability in that once the promise of continued upward mobility and prosperity waned, so too might that social contract erode.

Could technology then be wielded, as many China watchers openly wondered, to ward off that prospect? Maybe. The very push toward these frontier technologies seemed to be creating a magnetic effect for the world's top scientists, and what they might build on behalf of their Chinese benefactors. Indeed, the goal was to make China a "pacemaker" in high-energy physics, explained Wang Yifang, director of the Institute of High Energy Physics in Beijing, drawing in talent in much the same way America had in the twentieth century.

Still, as its ambitions grew and new labs and funding developed in the East, the existing work at CERN, as well as other national laboratories and commercial ventures, could not be discounted. CERN was at least for now a preeminent powerhouse for particle physics.

Besides, there was something already here that I needed to explore—the whisper of the kind of energy which, if ever harnessed, could reshape our very notions of human-made power.

# THE ANTIMATTER FACTORY

Even from a distance, the large blue sign with bold white letters at CERN has a distinctly sci-fi feel: more futuristic outpost than working laboratory.

ANTIMATTER FACTORY, it reads.

Push open the metallic door and the atmosphere becomes cool and sterile. Radiation warning signs flank the walls of the facility, filled as it is with the hum of machinery and white-coated physicists, including Dr. Elise Wursten, who would greet me there, at

With the particle physicist Elise Wursten, at CERN's Antimatter Factory in Prévessin-Moëns, France, July 2022.

the entrance of CERN's Antiproton Decelerator Hall, a facility that began operations in 2000. Inside was the kind of place where particles were smashed, antimatter was made, and Dr. Wursten seemed to quietly wonder what I was doing there. Here, scientific talent had met funding, and the research had thrived in a way that the U.S. no longer seemed to reliably support.

In truth, the building was used to create low-energy antiprotons by firing proton beams into heavy-metal blocks, such as tungsten or gold, which then created a cascade of secondary particles, not unlike other high-energy collisions studied elsewhere at CERN. The difference was that here they were slowed down. Captured and cooled, and electromagnetically suspended and trapped in a vacuum chamber, lest these antiparticles annihilate themselves in a burst of energy upon contact with matter. Again, it seemed like the stuff of science fiction. And yet, somewhat miraculously, that trap is precisely what Elise was now holding in her hand—a thermos-sized device to store antimatter, invisible to the naked eye, suspended in a sealed vacuum, and lethally delicate, like impossibly powerful fireflies confined in a jar. Just a single gram of the stuff could generate a whopping 90 terajoules of energy, equivalent to 21 kilotons of TNT or roughly the explosive power of a small atomic bomb.

"Can I hold it?" I asked.

"Yes," Elise replied. "This is the real one that was used."

She then placed the Penning trap into my hand. Named after the Dutch physicist Frans Penning for his work in ion trapping and plasma physics, the copper-colored, cylindrical trap was about the length of a ruler. Its metallic housing encased a series of coils wrapped inside a protective barrier, which encased the vacuum chamber that would confine the antimatter.

"This is just about the craziest thing I've ever held."

Elise laughed and then promptly took it back.

Scientists at CERN had, in fact, set a record for storing antimatter in the trap for 405 days, effectively demonstrating that it could be housed with relative stability. Storing portably, it held all sorts of promise, while also fueling ideas from the formulation of the universe to its dreamed-up use in frontier propulsion. There were, of course, major impediments. Given the immensity of the energy requirements of containment, as well as the extreme volatility of the particles due to instant annihilation upon any contact with matter—which could lead to devastating consequences if either the magnetic field or the vacuum conditions failed—storing larger quantities of antimatter was problematic, to say the least. Containment was a hurdle, to be sure, but perhaps not as much as that of sheer supply. Despite its name, CERN's so-called Antimatter Factory could produce only a few nanograms of antimatter at any given time, or 0.000000001 grams, hardly the sort of fuel source that could power the equivalent of the USS *Enterprise,* even *if* such an energy source could ever be harnessed and directed. It seemed far more useful in particle physics, cosmological research, materials sciences, or even medical imaging, where positron emission tomography scans are applied as diagnostic tools for detecting potential cancer clusters in the body.

Its potential in space propulsion, however, was something physicist Eugen Sänger had considered just decades after antimatter itself was predicted in 1928 by the Nobel Prize–winning British physicist Paul Dirac. Born two decades earlier, in 1905, in what was then Austria-Hungary, Sänger as a young man had drawn inspiration for interplanetary travel after reading *The Rocket into Planetary Space* by Hermann Oberth, who was born in what was then Austria-Hungary and would go on to help develop the Nazi V-2 ballistic missile program and mentor Wernher von Braun. Oberth himself was also said to have grown enamored with the possibility of space travel after reading Jules Verne's science fiction novel *From the Earth to the Moon,* and would later devise formulations that would become foundational to modern rocketry and spaceflight.

Sänger, in many ways, seemed to want to pick up where Oberth left off, and wrote his own book, *Rocket Flight Engineering,* in 1933, while proposing hypersonic craft and devising prototypes for liquid-fueled rocket engines and later ramjet engines, which could outperform turbine-based jet designs and be employed to reach supersonic speeds. Three years later, as Adolf Hitler sent German troops to reoccupy the Rhineland, which had been demilitarized following the Treaty of Versailles, German high command seemed to recognize Sänger's talents, later evaluating his designs in building a secret space bomber called Silbervogel (German for "Silver Bird"), also known as the Amerika Bomber. A suborbital crewed vehicle meant to skip along the edges of space could perhaps provide a Nazi capacity to attack distant American cities like New York. But like many advanced Nazi projects, the supersonic, stratospheric plane would be developed too late to affect the war, its designs nonetheless influencing later concepts like America's X-15 hypersonic rocket plane, as well as early designs of the space shuttle.

And yet it was those earlier designs for a photon rocket, nearly half a century before CERN's unique Antimatter factory began producing antiprotons, that devised plans for employing gamma rays generated by the annihilation of antimatter for propulsion that were still being considered today.

In a paper published in 2017, in the peer-reviewed scientific journal *Acta Astronautica,* a Norwegian professor named Dr. Espen Gaarder Haug argued that not only was a photon rocket theoretically possible, but it offered the potential for maximum propulsion that approached fractions of light speed. His contention unsurprisingly drew skepticism, especially given his unconventional background. Haug, a professor of quantitative finance at the Norwegian University of Life Sciences, argued that there are mathematical corollaries in his field applicable to physics, after having written about the potential use of neutrinos produced by particle accelerators in high-frequency trading. Others within the scientific community lauded his work as having genuine merit, having been published in peer-reviewed, physics-related journals. Still, he acknowledged that "there is still a long way to go in developing large photon rockets that could send materials or people into outer space," especially given the sheer limitations on the amounts of positrons that exist or could be created.

The amount produced thus far at CERN was barely enough to power a lightbulb, let alone the vast quantities needed for an interstellar journey. What's more, physicists and engineers would have to keep the rocket itself from being destroyed, given the power of the 100 percent antimatter-to-matter energy conversion. A cloud of electron gas at the rear of the craft, Sänger postulated, might be adequate to reflect the gamma rays produced in that positron-electron annihilation, which could theoretically, he supposed, power the craft at mind-bending speeds. But again, how precisely that gas might be achieved was not clear.

Other concepts, meanwhile, included methods like antimatter-catalyzed nuclear pulse propulsion, in which antiprotons would be injected into a nuclear fuel source to kick-start a chain reaction, which then could theoretically drive electric propulsion engines. Unlike the more direct positron-electron annihilation approach, that antimatter would instead power more traditional systems like ion or Hall thrusters. It was all big-picture stuff, sci-fi ideas that were often treated as such. Still, curiously, progress was being made.

Even if the problems of physics and engineering could be overcome, however, the vast distances between stars would create problems ranging from extreme communications lags to profound psychological effects tied to prolonged isolation, confinement, and a general detachment from Earth, not to mention problems like time dilation, given interstellar voyagers would experience the passage of time at different rates from those they left behind, a phenomenon already evidenced by time discrepancies with satellites in Earth's orbit. Given those considerations, it begged a deeper, perhaps more all-encompassing question: Did the pathways for true interstellar travel require a fundamental rethinking altogether?

In probing such frontiers-of-physics ideas, the trail again turned south, only this time to Mexico City.

· CHAPTER 45 ·

# WARP DRIVES

When I first phoned Miguel Alcubierre, he made one thing clear before we started talking. "What I really do normally is study black holes and gravitational waves." Born in Mexico City, he had earned a PhD from the University of Wales, centering on numerical general relativity, which employs supercomputers to analyze phenomena described by Einstein's theories, such as black holes, neutron stars, and gravitational waves. His more recent work focused on the study of orbiting black holes that spiral into each other and collide in an outburst so powerful that they cause ripples across space-time, similar to a stone being cast into a still lake. The field was meant to offer insights into theoretical marvels like quantum gravity, the nature of dark matter, and even more exotic theoretical concepts like wormholes, which had been predicted through general relativity and even at times considered by those on the scientific edges as potential shortcuts through the fabric of space.

But that's not what I was calling to discuss.

Rather, it was his work as a graduate student in Cardiff, where as a young theoretical physicist he would make a discovery that would later bear his name.

Ironically, it all started with *Star Trek*.

The sci-fi series *The Next Generation* ran from 1987 to 1994, depicting fictional interstellar voyages across galaxies. Miguel was a fan. And one evening, he slid in a VHS tape to watch an episode. The show's opening sequence illustrates a spaceship moving through distant nebulae and ringed planets, promising "to boldly go where no one has gone before." Then, the ship surges forward, seemingly warping space around it as it vanishes into the distance. The idea intrigued him. And as he settled in to watch, Miguel found his thoughts drifting to the idea of a warp drive. "Traveling faster than light isn't possible," he said. "But they do it all the time on TV."

"They talk about warp drives, but they never explain what they actually are—because, of course, it's just science fiction," he added with a shrug. Speed, of course, did have its limitations, as were clearly defined by relativity. Yet intriguingly, there also appeared to be loopholes: a theoretical exception that somehow did not violate any known laws of physics. Space itself, it turns out, is not bound by any such speed limits. That meant the fabric of space can, in fact, expand faster than light—a concept intrinsic to cosmic inflation theory.

To illustrate his thinking, Miguel made the following analogy: Take a rubber band and use a pen to mark a series of dots along it. Each dot represents a galaxy. Now stretch the band. The dots move away from one another, not because they're traveling on their own, but because the band itself is expanding. This simple act demonstrates a fundamental concept: While galaxies have their own individual motions, they aren't flying apart through space so much as space itself is stretching between them. Plus, the farther away a galaxy is, the faster it appears to be receding—which suggests that at far enough distances some galaxies are receding faster than the speed of light. The very expansion of the universe itself demonstrates that. Hubble's law, in fact, states that galaxies move away from Earth at speeds directly related to their distance—which is referred to as the cosmological horizon, or the maximum distance from which light could have traveled. That essentially constitutes the very boundaries between the observable and the unobservable universe, which means there are parts expanding so quickly that light cannot catch up.

Given that, he surmised, "I was just thinking maybe there's a way to allow you to do what they claim to be doing in science fiction," Miguel explained. All of this was apparently swirling around his mind as he watched *Star Trek,* whose creator, Gene Roddenberry, had a clear fascination with faster-than-light travel.

"Warp drive, Mr. Scott," bellowed Captain Kirk, played by William Shatner, in the original series. "Accelerating to Warp One, sir." Stars then turn into dashes as the USS *Enterprise* disappears into a dot against the blackness of space.

The concept had Miguel hooked. And by looking at cosmic inflation, he figured that the same idea could be used to devise a warp drive, or at least create a believable theory for one.

"I had this idea," he explained. But "I don't like the term 'drive,' even though I use it," he added. "It's more of a bubble."

Given space-time is often described as a fabric, massive objects can create indentations in it. Much the way a bowling ball, if placed in the middle of a bed, creates a dent in a bedsheet, if a powerful gravitational force could be generated and combined with exotic matter that held negative energy properties, it might then be possible to manipulate space-time itself. By essentially contracting space in front of a spaceship and expanding it behind, a warp bubble might then theoretically form. Inside that bubble, the spacecraft would, in essence, remain stationary, just as the bubble carried it across vast distances at faster-than-light speeds.

Think of it this way—a person standing on an airport conveyor belt moves even if they do not take a step.

There is a catch, however. The arrangement required far more energy than anything humans could possibly produce with today's technology, as well as this mysterious concept of negative energy density—which means an energy less than zero. That's something that doesn't exist in the classical sense. And yet the mere intellectual exercise of it all had enlivened Miguel.

"The next morning, I was in my office at seven doing calculations to try to figure out if this thing made any sense," he explained. "If I wrote down a geometry for space and time, that would allow me to express this idea in mathematics. I was there for several hours. And eventually, yes, I came up with that, with a simple geometry that contained this basic idea." The next challenge was to figure out precisely the level of energy required, plus this concept of negative energy, which Einstein's relativity allowed for but remained highly speculative.

"I came up with the fact that they needed negative energy," Miguel explained, given it would need some sort of repulsive effect to work against gravity and allow for a manipulation of space-time without collapsing the bubble.

"Initially, I did it in two dimensions, one dimension of space and one dimension of time. And then, once that worked, I moved to four dimensions, which is what we actually have: three dimensions of space and one dimension of time."

As he toiled, the ideas began to take firmer shape.

"Eventually, I showed the notes to my PhD adviser, and I was worried that he would say, 'This is completely ridiculous; go back to work on black holes.' But he actually read it and he said, 'Miguel, this is really interesting. You should publish it.' And I said, 'I've never published anything in my life.' And he said, 'I'll help you.' "

At that moment, listening to Miguel's explanation over the phone, I felt I had to interrupt.

"Wait," I said, interjecting. "This was the first paper you ever published?"

"Yes," Miguel replied. "He also helped me with the title."

They needed something bold and intriguing. What they landed on would have made Captain Picard proud: "The Warp Drive: Hyper-Fast Travel Within General Relativity."

In it, Miguel explained how, even without the use of wormholes to carve out cosmic shortcuts, physics nonetheless provided a theoretical means to modify space-time "in a way that allows a spaceship to travel with an arbitrarily large speed."

It was a bold claim, one that seemed at least initially to dance on the edge of speculation and scientific rigor. And, of course, it was all controversial, inviting both excitement and skepticism, not least of which centers on something known as causality.

In causality as we understand it, events unfold in a linear sequence. An apple falls only after it is dropped. Faster-than-light travel, however, could undermine causality, since signals or objects would arrive at a destination before they were sent. Such an idea is deeply unnerving, in that it would effectively violate the fundamental cause-and-effect relationships that underpin physics. Violating causality would also open the door to paradoxes about our very notions of reality, in which actions today could alter events in the past. That is largely why the principles of causality remain within Einstein's theories of relativity. In both special and general theories of relativity, the German-born physicist effectively preserved causality as essential to nature itself.

Yet even Einstein allowed for exceptions in the case of space-time expansion. In this framework, it's not the spacecraft that moves faster than light; rather, the space-time around it is being manipulated. In this case, Alcubierre's concept would apparently not violate causality. The idea certainly involved faster-than-light travel, but that travel would occur only within a localized "bubble" that moves through space-time. Crucially, within that bubble, no information or matter would exceed the speed of light relative to the bubble's internal frame of reference. Causality, in this sense, would presumably be preserved.

Questions remained, however, about whether this preservation would hold up from a more global frame of reference. In other words, could observers outside the bubble witness violations of causality that wouldn't be apparent from within? And beyond that, what unforeseen consequences might arise from the extreme warping and manipulation of space-time itself?

These were all wildly unsettling questions. Then again, at the time, they were derived from little more than theoretical exercises. There was nothing in Miguel's theories that could be operationalized, let alone employed to devise construction plans for the USS *Enterprise*.

"That was back in '94, and in the course of the three decades that have elapsed since then, there have been all sorts of iterations," he said.

Initially, it appeared that such an experiment in practice would require "as much negative energy as that that existed in the entire universe," which seemed more of a qualitative way of expressing impracticality than it was a precise measurement.

It all, therefore, ultimately just seemed like a fun experiment for future physicists to ponder. Yet over time, those very theoretical foundations began to shift. New breakthroughs in quantum field theory and exotic matter research would advance the concept. Later iterations of his ideas would, for instance, demand considerably smaller, though still unobtainable, energy requirements, tantamount say to the mass of Jupiter. Thicker walls of the accompanying theoretical structure could theoretically also decrease those energy requirements. But still, it wasn't scalable. Thicker and thicker walls, in other words, did not produce lower and lower energy requirements.

In fact, Miguel added, "if the walls become very thick, you have no bubble." Plus, "negative energy simply does not exist as far as we understand the laws of nature today," he added, referring to its lack of existence in classical physics. The broader idea still seemed relegated to realms of pure theory. And yet his paper had had a galvanizing effect. What had once been viewed as pure science fiction suddenly edged the scientific community ever so slightly closer to realms of plausibility, with many now openly wondering whether the principles of relativity and causality could be bent without breaking.

· CHAPTER 46 ·

# CASIMIR

In the pantheon of the physics community, the twentieth century stands out as a remarkable era of notables. Those like Albert Einstein would become household names. Others, like Niels Bohr, who helped decipher rules of quantum mechanics, and Paul Dirac, who predicted antimatter, are generally celebrated more within the confines of the communities they served. And yet the name Hendrik Brugt Gerhard Casimir doesn't quite get the spotlight it deserves. To his colleagues, including the Dutch physicist Dirk Polder, Casimir, with his "powerful intellect and exceptional memory, his strong associative imagination and great feeling for language, and his generous spirit, backed up by a robust physique . . . was destined to play a dominant role in the world of science."

But it almost didn't happen. The classics indeed were calling. In fact, Casimir's decision to fully forgo a career in classical literature to pursue the atomic and subatomic worlds was apparently only really decided after his adviser, Paul Ehrenfest, brought him to a 1929 conference in Copenhagen where he would meet Niels Bohr, then considered among the most influential figures in modern physics.

"Now you are going to make the acquaintance of Niels Bohr and that is the most important event in the life of a young physicist." To underline the point, in his introduction to Bohr, Ehrenfest said, "I am bringing you this boy—he has some abilities but he still needs thrashing."

Hence began the career of young Casimir. As his work matured, he moved to Eindhoven and began examining what appeared to be minuscule fluctuations of energy that emanated not from matter but from the void itself. Even in empty space, fluctuations at the quantum level can have observable consequences, resulting in something called negative energy density in regions where fluctuations are more restricted—like in the space between two closely positioned metal plates.

This is what Miguel had homed in on. To manipulate space-time through expansion and contraction, a unique form of energy was needed—one characterized by values that were head-scratchingly less than zero. But what did that even mean? Energy was generally considered to have positive value, at least in the classical sense. We know how to account for 1 joule of energy. But how does one account for -1 joule? To quote the physicist John Wheeler, "If you are not completely confused by quantum mechanics, you do not understand it."

While the concepts of negative energy densities exist in physics, they were generally not observable in nature, except in a few arenas of quantum theory, such as those found in Casimir's early observations. Hence known as the Casimir effect, the work that was based on those early observations would offer tantalizing demonstrations of the kinds of forces that might be needed to one day make Alcubierre's warp drive appear more plausible.

Sure, there was nowhere near the amount of it currently attainable that could fulfill the warp bubble's energy requirements. But if those energy requirements had already proven that they could be reduced, could they one day be reduced even further so as to breathe life into what was so far just a purely theoretical concept? Many were skeptical. And yet in these new AI- and quantum-fueled realms of science, innovations were being propelled forward at a dizzying pace.

But first, let's take a step back to evaluate what was actually being contemplated.

To better illustrate the concept, picture this: a lone ring-shaped island in the ocean during a storm. On the outside, strong winds and waves pound the shorelines. But inside, where coral reefs create a protective barrier, the waters are calm. In particle physics, that contrast between the calm and the stormy generates a pressure difference, or a difference in energy densities. As a result, the energy inside can exert a negative effect on the energy outside, or a negative energy that stems from the difference caused by the two parallel plates.

With the right architecture, could that energy then be harvested out of what appeared to be sheer nothingness? Any device for extracting Casimir energy would still have to adhere to the laws of thermodynamics, which severely limit its application, given the expenditures of energy presumably outweigh the possible energy that could be extracted. Sure, the Casimir effect could impact vacuum energy configurations and even produce measurable forces, many physicists I interviewed pointed out, but what it does not allow for is the harvesting of free energy in a way that results in net positive energy output without comparably external energy input. In other words, you're not getting any more out of it than you put in. Nonetheless, it was a concept that commonly intrigued graduate students given that it still hinted, however faintly, at the possibility of tapping into a latent power source within the quantum vacuum itself.

Meanwhile, emerging research exploring subtle anomalies and theoretical loopholes, while also exploiting new advances in fields like nanofabrication, seemed to cast light on related quantum concepts. The field itself seemed poised for explosive growth.

By 2008, for instance, Masahiro Hotta, a theoretical physicist at Tohoku University in Japan, suggested that energy could be transferred (or teleported) in the quantum state. Hotta essentially proposed that by exploiting fluctuations, one could possibly transfer energy using the properties of entanglement and superposition, two core features of quantum mechanics, to enable what was tantamount to a non-local transfer of energy between distant points in space, without the energy ever physically having to travel through the space in the classical sense. This raised provocative new questions about the nature of energy itself.

Physicists Richard Obousy and Aram Saharian would later argue, "In principle, it may be possible to directly manipulate the size of an extra dimension locally using Standard Model fields in the next generation of particle accelerators." The paper also pointed to the suggestion "that the Casimir effect in higher dimensions may be the underlying phenomenon that is responsible for the dark energy which is currently driving the accelerated expansion of the universe." The authors noted that it may therefore be possible "to locally adjust the dark energy density and change the local expansion of spacetime," which would then hold "tantalizing possibilities in the context of exotic spacecraft propulsion."

It was an intriguing idea, though, again, still held to the confines of pure theory and considerable skepticism.

Then, in 2011, a former NASA physicist named Harold "Sonny" White sought to reignite interest in Alcubierre's faster-than-light travel concept when he published a provocative paper suggesting that the Casimir effect might be employed to create the kinds of negative energy densities required to generate a nanoscale warp bubble. It was a bold and controversial refinement of Miguel's original concept, suggesting theoretical notions of warp-speed travel could find support in the physics of the quantum vacuum. Many considered it unworkable both in unresolved engineering hurdles and the fact that it would likely violate known thermodynamic constraints. And yet the tantalizing possibilities of a warp bubble, even if they were at the fringes of physics, also seemed too enticing to merely discard without further investigation.

This was all a lot to take in. Time-honored frameworks were now intersecting with fresh speculations, increasingly powered by a wave of new technologies that were only just starting to come online.

So, in a bid to get a better sense of it all, I traveled to Texas to meet with Sonny in person. We met at his lab in Houston, at a Kam Ghaffarian–funded organization dedicated to interstellar innovations and STEM education. Sonny, who had worked as advanced propulsion lead for NASA's Engineering Directorate and served as the Johnson Space Center representative to the Nuclear Systems Working Group, had focused on advanced power and propulsion technologies for much of his career, and now was intent on explaining the intricacies of what his lab was exploring. Our first meeting lasted for more than four hours. Everything from negative energy densities to warp bubble concepts were on offer, with equations eventually littering the whiteboard behind him, and Sonny's occasional look-back to make sure that I followed.

The equations, of course, didn't help. Not for me at least. In fact, to me, this counterintuitive quantum world initially really just seemed like the bedrock of little more than science fiction. Increasingly, however, it grew more fascinating as he spoke, albeit with some important caveats. His paper, titled "Warp Field Mechanics 101," was not published in a peer-reviewed journal but instead presented as part of NASA's Advanced Propulsion Physics Laboratory's research—thus considered more framework than empirical study. And yet the paper seemed to harken to the potential for a field in transition, noting that while "warp field mechanics has not had a 'Chicago Pile' moment, the tools necessary to detect a modest instance of the phenomenon are near at hand." ("Chicago Pile" is a reference to the world's first artificial nuclear reactor.) Those tools included increasingly sophisticated nanofabricators and the growing potential of AI, which Sonny seemed intent on leveraging.

"I did not want to just parrot what has already been done," he told me. "Rather, I decided to do a sensitivity analysis of the field equations to see about reducing the energy density requirements. My toy model was a ten-meter-diameter spacecraft with an effective velocity of 10c [shorthand for a velocity of ten times the speed of light]. I wanted to duplicate the Jupiter amount of exotic matter calculation done by Dr. Richard Obousy, which arises as a result of a very thin ring of negative energy density around the craft. This thin ring causes the York time magnitude to be colossal. York time is like a three-dimensional strain.

"In my sensitivity analysis, once I had my starting point," he added, "I increased the ring thickness more and more; further, I oscillated the bubble intensity, which serves to reduce the stiffness of space-time. The end result was an estimate of the negative energy density that was less than the mass of the Voyager 1 spacecraft.

"This," he concluded, "at least moves the idea from impossible to plausible."

I had no way knowing if that were true. But the paper itself went on to describe the Alcubierre concept in practical terms as it relates to a hypothetical interstellar spacecraft, noting the following:

> The field would be turned on and the craft would zip off to its stellar destination, never locally breaking the speed of light, but covering the distance in an arbitrarily short time period of time [*sic*] just the same. The field would be turned off a similar standoff distance from the destination, and the craft would finish the journey conventionally. This approach would allow a journey to say Alpha Centauri as measured by an Earth-bound observer (and spacecraft clocks) measured in weeks or months, rather than decades or centuries.

"Feasible," he later added, "is still future work."

Sonny's approach was effectively to try to reduce the overall reliance on negative energy, by improving field efficiencies and the overall distribution of energy density of the bubble, in part by altering the shape of the bubble itself.

"It's mathematically possible," he told me in front of that whiteboard, again filling it with equations in that Houston office. "But no one can stand in front of you today and say, 'Here's the assembly documents for the USS *Enterprise*. Go take this to Newport Shipyard, and they'll have this thing for you in ten years, and you can go do whatever you want with it.' Right?"

Even *if* plausible, however, there was a mountain of challenges, namely the fact that Casimir effects could potentially produce only minuscule amounts of negative vacuum energy, nowhere even in the vicinity of the amounts required, even with Sonny's proposed energy reductions. Still, Casimir had proven an exciting source discussion amid an ever more power-hungry human society. Given that quantum fluctuations in a vacuum between two closely spaced conductive plates could produce tiny but measurable forces, what if those plates could be stacked and scaled? There were plenty of reasons why it couldn't be done, including the extraordinarily weak effects of its force, the engineering constraints of stacking thousands of plates in a vacuum, and those more foundational critiques that question whether energies were indeed something that could be continuously drawn upon. There were even emerging questions as to whether these particular energies were derived from the vacuum at all.

Still, the discussion was hard to ignore. Sonny, meanwhile, was eyeing the creation of small-scale computational modeling to decipher feasibility. Again, the idea was steeped in controversy, drawing suspicions even from Miguel himself, given Sonny's warp drive paper was not considered an empirical study.

"I simply don't believe the claims that Sonny White has been making," Miguel later told me. "He didn't create anything in the laboratory, because the paper is just a computational simulation. . . . Yes, the Casimir effect is real. When you put parallel plates with different geometries, we can in principle create a region of negative energy inside those places. But the amount of it is so small."

Nevertheless, curiosities across the physics landscape seemed reignited about the very boundaries of what was possible, including questions about precisely what Sonny White would later nanoconstruct in his lab, and what applications these new microchips might suggest. Could the quantum field ever be applied for more practical energy needs, not just in deep space, but here on Earth?

The answers were now being roundly debated. And perhaps that was the point. Pursuing the stars held the potential of unlocking new possibilities on our own planet, just as it had during the days of Apollo. Exploration for its own sake could essentially help illuminate the very edges of science, and in the process create new value propositions, even if they stumbled in the beginning, or if the merits of the efforts were questioned altogether.

Meanwhile, alternative approaches to interstellar travel options were being explored elsewhere. In Switzerland, for example, researchers were working on devising new methods, perhaps even without the need for exotic matter. In a 2024 study, published in the peer-reviewed scientific journal *Classical and Quantum Gravity*, the lead author, Jared Fuchs, a senior scientist at the research firm Applied Physics, said he sought to change "the conversation about warp drives" with the firm's own theoretical modeling. Titled "Constant Velocity Physical Warp Drive Solution," the research involved "combining a stable matter shell with a shift vector distribution that closely matches well-known warp drive solutions such as the Alcubierre metric."

It was but a single study, and it was far from clear whether a warp drive could ever be created, let alone employed, to make humans a galaxy-hopping species. But the effort was gaining traction, including among those in Beijing.

The future now seemed open for the taking.

• PART IV •

# CHANGE

· CHAPTER 47 ·

# ADAPTATION

Think of a bacterium: small, vulnerable, and yet highly adaptable. A single strain can restructure itself in just hours. In the veritable jungle that is the human gut, survival often depends on it. Now think of an elephant. In South Africa, researchers uncovered a remarkable phenomenon. After generations of poaching pressure over highly prized ivory tusks, many are now born without them. Of the 174 females in Addo Elephant National Park in 2000, a full 98 percent were tuskless at birth.

Life is indeed the story of selection and change. Almost every cell of every organism has a regulatory process that feeds transformation, adjusts for new environmental conditions, or is simply reflective of changing realities. For some, like the *E. coli* in our gut, it can happen quickly. For others like the elephants of Addo, it is multigenerational, or, as Charles Darwin described it, marked by "the long lapse of ages."

Life adapts. Then again, species' individual resilience is also hardly guaranteed. The Huronian ice ages preceded several mass extinctions. The Cryogenian period plunged Earth into its deepest, coldest icy hell. In fact, over the last half billion years, changes in climate and ocean chemistry have edged life to the very brink of disappearance, with an estimated 99 percent of all species wiped clean from the Earth. At times, it was only unicellular microbes, clinging to life along frozen seabeds. At others, more complex creatures, such as dinosaurs, roamed the planet. Each had their moment, though certainly not on equal terms. Horseshoe crabs, with their myriad adaptations that afford them a capacity to live across a variety of habitats, have survived across more than 480 million years of history. Others, vulnerable to rivals and reliant on more fixed locales, such as the dodo, died out only two hundred years after their discovery.

The jury is still out into which category we will fall. It's also unclear which system could best secure our own longevity. But authoritarian control, be it government run or corporate influenced, seems very much at odds with free-thinking societies, especially given the rate of change over how we actually classify intelligence, be it human or otherwise. Either way, the time to decide the orientation of our future is fast approaching. We ought to be intentional, given history is not often kind to those who succumb to the inertia of unfavorable circumstances.

For thousands of years, *Homo sapiens* have lived without rival or relative. Yet just last century, sometime around 1950, Earth is believed to have entered the Anthropocene epoch—an unofficial geological time interval when human activity became so profound that its global effects began influencing our planet's natural systems, including atmosphere, oceans, and broader processes. In the past, hallmarks of such shifts have been preserved in rock layers and included changes in ocean chemistry, alterations in climate patterns, and mass extinctions. Today, even a brief survey of the planet's situation suggests something similar.

The assumption, of course, may be that humans will again adapt, given our species has proven remarkably resilient over our relatively short existence, and thus afford us horseshoe-crab-like staying power. And yet given how new we all are, it's hard to know.

As rapidly advancing technologies converge, we find ourselves at the threshold of a reshaping human society, unfolding at a pace faster than we can comprehend and supported by emerging economies, power struggles, and technologies in the arena of space. If we are to survive and figure this out, we're going to need to work together. And yet presently, our politics is an impediment.

Perhaps, that—not the tech—is then the deeper question.

As I mentioned at the start of this book, I have every confidence in humanity's ability to engineer just about anything. But what we should consider more urgently is our ability to organize ourselves and steward a future with the kinds of systems that produce leadership attuned to this radical transformation. The stakes are existentially high, not merely as we press on to new celestial locales, but in a broader struggle for who and what we will be.

To uncover real answers and lasting solutions, at least one thing seems abundantly clear: The East and West need not be enemies. Our perspectives and visions may differ, and we'd be foolish not to fortify ourselves with strength and sustainability in such uncertain times, given that weakness is a clear invitation for exploitation. But that difference also need not lead to conflict. Lasting influence is, in fact, hardly ever a product of hard power alone. If we can remember that, and perhaps use these emerging platforms on the Moon and in low Earth orbit for a bit of perspective, perhaps we can also gain the sort of wisdom that astronauts who have gazed upon our planet from afar have long since described.

That is, the Earth is fragile, our destinies are interconnected, and our only real chance at navigating this new reality is not only by thinking strategically in the context of competition, but also by listening to each other, recognizing shared threats, and finding meaningful ways to work together.

# EPILOGUE

The idea for this book came in a few different forms, ranging from an upbringing with a mother who wanted to be an astronaut, the enthusiasm of a literary agent involved in the 1986 Voyager flight, and those early, undeniable headlines about SpaceX, Blue Origin, and China, which all seemed to hint at a tectonic shift in the commercial and geopolitical relevance of space. Yet it also came from a trip to an icy region situated halfway between Norway and the North Pole, during a documentary filmmaking expedition in the fall of 2014.

Svalbard is sort of a blue twilight world, where ships slalom icebergs, polar bears outnumber people, and the midnight Sun gives way to months of darkness. At 78° north, this wilderness of labyrinthine fjords and snowcapped peaks is juxtaposed to brightly colored wooden buildings that make up its main settlement of Longyearbyen. But it is the archipelago's low light pollution and northern latitude that make it particularly well suited for the observing of space, especially when it comes to the aurora borealis.

To the naked eye, these highly charged particles—streaming from the Sun—can be seen racing to the poles and shedding energy in the form of light. And on that particular trip, having been woken up by a 3:00 a.m. phone call, I was lucky enough to catch a glimpse of this natural phenomenon. Rivers of emerald, crimson, and magenta twist and flow overhead. And for whatever reason, my mind then drifted to what it must have been like for a more ancient people, gazing up as they must have done. Had they devised legends and mythologies as a means of connecting this celestial order to a more relatable human experience?

Stargazing, to be sure, pairs well with a good story. It was something that would stick with me, especially when I became a father.

As adults, we often look down—at our children, at our work, or in fumbling with our phones. But when we look up, as our children do, it is hard not to share in their same sense of wonderment, especially in places that reveal the glowing tapestry of stars that surrounds us. In fact, there may even be something about Earth's remote locales that influence those who look to space. John Glenn—the first American to orbit Earth—and Neil Armstrong—the first man on the Moon—both hailed from rural midwestern towns. Yuri Gagarin—the first man in space—grew up in a small village. Travel beyond urban centers, and past the artificial glows that can wash out the night's sky, and "you can know light," as Janet Kavandi, a NASA veteran of three space shuttle missions and former president of Sierra Space, explained to me, while reflecting on her own childhood memories of stargazing in Jasper County, Missouri, with her father.

"I now have a house in Houston," she explained. "And I can't see anything."

That sort of account, it turns out, is remarkably common among those devoted to the cosmos. The CEO and founder of Rocket Lab, Peter Beck, who grew up in one of the world's southernmost locales, told me a similar story.

"One of the youngest childhood memories I have was my father taking me outside," he explained. "We lived in a small town [in New Zealand], and you could see the night sky and Milky Way. And I remember him vividly pointing to the stars and saying, 'Those are stars, and around those stars are planets. And there could be somebody looking back at you.'

"That, for me, absolutely blew my mind." Those moments, beyond the diluting reach of urban light, were "really the beginning of my whole career and love for space," he told me.

Hearing each of them speak, I couldn't help but reflect on my own travels, changes of locale, and what experiences I wanted for my own child. Perhaps that was the pull that urged me to leave New York City for the red rocks of Arizona and its starlit black skies. The idea was not just to pursue Earth's wild places, but to show them to my daughter and thus share in her sense of awe, even if, at the time, I didn't quite understand why.

# Special Acknowledgments

*A special thanks to Peter Riva and Victoria Wilson, without whom this book would never have been written.*

*Thank you also to Brewster Shaw, Steve Altemus, Josh Marshall, Steven Hirshorn, Charles Miller, Ali Velshi, Ling Xin, Cat Hofacker, Doug Loverro, Bettina Inclán, Quynh Do, Margot Lee, Scott McGhee, Teall Gerrett, Katie Walmsley, Nicole Wong, Antony Michels, Wyatt Potteiger, Tobias Bloyd, Simona Kraberger, and Neil Hallsworth, whose insights, friendship, and exchange of ideas proved indispensable.*

*Thank you especially to my father, Robert, my mother, Deborah, and my sister, Elizabeth, as well as my entire family, whose love, support, and humor I could scarcely do without.*

*Finally, thank you to Rose, who—among other things—reminded me to look up at the stars and tell stories, just as our ancestors did . . .*

*. . . even if many of those stories are not yet fully written.*

CLPS
COMMERCIAL LUNAR
NASA
NOVA-C
IM-1
ADTIGO PLANITIA LUNAE

# Notes

### CHAPTER 1: Launch Day

13 "Decisive action will demonstrate": Kam Ghaffarian, "The Trump Administration Should Leverage Private Space Stations to Counter China," *SpaceNews,* Jan. 13, 2025.

15 Three decades later, he had amassed: Similar to Musk and Bezos, however, Kam seemed to have a penchant for those bigger, society-shaping ideas. For Bezos, calls for a "sustained human presence," and the moving of heavy industry off Earth, seem to complement his own ideas for human space populations to live in so-called O'Neill cylinders—massive rotating chambers proposed by the physicist Gerard K. O'Neill in his 1976 book, *The High Frontier: Human Colonies in Space*. These large, free-floating space habitats were to be theoretically constructed using resources extracted from the Moon and asteroids as a way of alleviating resource and population burdens on Earth. They were—again theoretically—intended to house as much as 90 percent of future human populations, a somewhat fantastical notion emblematic of an early space era when humans dreamed big. In fact, as a precondition to being interviewed about his space company, Blue Origin, the Amazon founder has reportedly asked journalists to watch a 1975 PBS program that featured O'Neill and Isaac Asimov. Asimov had been considered among the "Big Three" science fiction writers, alongside Arthur C. Clarke and Robert Heinlein. Musk, in fact, had often cited Asimov's sci-fi series *Foundation,* which depicts the collapse and return of an interstellar empire, as both galvanizing his own interest in space and foreshadowing societies' cyclical rise and fall. In several interviews, the South African–born tech billionaire outlined the importance of taking a "set of actions that are likely to prolong civilization, minimize the probability of a dark age and reduce the length of a dark age if there is

one." He also frequently cited Douglas Adams's 1979 book, *The Hitchhiker's Guide to the Galaxy,* as a "philosophy book disguised as humor." SpaceX's first Mars craft, in fact, is to be named "Heart of Gold" as an homage to the ship in the novel. Central to his ambitions, Musk was busy "mapping out a game plan to get a million people to Mars," while the SpaceX Starship—the largest and most powerful rocket in history—would be "designed to traverse our entire solar system and beyond." Musk also added that "civilization only passes the so-called single-planet Great Filter when Mars can survive even if Earth supply ships stop coming." The implication was that intelligent civilizations tend to either self-destruct or succumb to external forces before they can attain multiple points of redundancy across the cosmos. That concept was articulated earlier in the 1990s by the economist Robin Hanson in a bid to explain not only why intelligent life is rare but that its sustainability might be even more unusual.

15 But unlike Elon Musk and Jeff Bezos: To be clear, Elon Musk's SpaceX was not his only space-oriented endeavor. Those like Neuralink and the Boring company also had clear space-related use cases. But his early focus seemed clearly oriented in getting those Falcon rockets, and later Starship, into space. Musk, having studied physics and economics, had no formal engineering degree. Yet he became SpaceX chief engineer, CTO, and CEO and remains heavily involved in engineering decisions at Tesla.

15 harsh realities of the Red Planet: Intuitive Machines' Nova-C lunar lander employed a mixture of liquid oxygen and liquid methane as a propellant, precisely the kind of rocket fuel likely to be developed in the Martian environment, which those like Musk envisioned for his megarocket, Starship. A process called the Sabatier reaction, invented by the French chemist Paul Sabatier, uses electrolysis and a mix of carbon dioxide (which humans breathe out) and hydrogen (which is found almost everywhere) to generate both methane (for rocket fuel) and drinkable water (for human consumption). SpaceX, Blue Origin, and United Launch Alliance are all developing methane-powered vehicles, as is the New Zealand–based Rocket Lab. But it would be the China-based LandSpace that would notch the first major win in 2023, sending the first methalox-fueled vehicle into Sun-synchronous orbit.

## CHAPTER 2: Building Landers

18 Having served as engineering lead: The precedent of Morpheus had been a problem. During testing, the craft had lost data and kept pitching until it smashed in a fiery blaze. To make matters worse, the entire event had been live streamed to the viewing public. The fallout was immediate. With its political support siphoned, the project lost funding at a time when watchdogs felt the agency was being pulled in "too many directions for too long." NASA, which its critics accused of becoming too risk averse and bureaucratic, just didn't seem ready for those finicky methane-powered, or methalox, engines.

18 The mix ignited on contact: "The old reliable" fuels were MMH (monomethyl hydrazine) and NTO (nitrogen tetroxide).

18 In commercial space: There, of course, were problems with methalox engines. Liquid methane boils off at a frigid -258.7°F, which means it requires extremely low temperatures to keep in liquid state lest it vaporize. Rocket-grade kerosene, by contrast, can be stored at room temperature. The first stage of the Saturn V rocket alone, which sent three Apollo astronauts on their way to the Moon, had carried enough kerosene (and liquid oxygen) to fill an Olympic-sized swimming pool. But more recently, NASA seemed to prefer using liquid hydrogen, which powered its Space Launch System's four RS-25 engines, which were ultimately meant to return Americans to the Moon. Part of the reason was its efficiency relative to kerosene. The Sun, as well as the gas giant Jupiter, for instance, both mostly comprise hydrogen. And measurements from NASA's New Horizons space probe found that our closest interstellar neighborhoods are in fact littered with hydrogen atoms. Future spacecraft would have comparably little trouble locating the element as a fuel source. Both were highly flammable, but hydrogen's lower ignition energy made for a far more efficient combustion. What's more, methane-powered engines at that time had never been used in deep space.

19 It was a key ingredient: Starship, in terms of Mars, was the real game changer. Capable of carrying up to 150 metric tons of cargo into orbit, or roughly the weight of a blue whale, its thirty-three methalox-guzzling Raptor engines were the most powerful launch vehicle ever built. Musk wanted to use it for Mars. And, at least initially with the Trump presidency, boosted by the cash and cachet as SpaceX founder, he seemed to get more of a chance. During the campaign, Trump routinely appeared with the SpaceX founder, who often sported his signature black OCCUPY

MARS T-shirt during rallies. Days before the election, the two men watched a Starship test flight together at SpaceX Starbase in South Texas, although relations seemed to have later soured between the two men. The company had initially planned to send five megarockets to the Red Planet by 2026, when Mars's and Earth's orbits were more closely aligned. If successful, crewed flights would then be expected in as soon as four years. SpaceX, NASA, and China were in fact all drumming up plans to send humans to the Red Planet, devise habitats, and eventually create settlements. The idea was to have recurring flights to build out colonies, from which broader developments could emerge, and perhaps even city-states might flourish.

20 The pushback in both cases: And yet the appetite for commercial activity, including space stations, was also growing. Commercially owned and operated low Earth orbit facilities were being directly supported by NASA's Commercial Low Earth Orbit Development Program for a host of companies, including Axiom Space, Starlab Space, and Blue Origin in partnership with Redwire, Sierra Space, and Boeing. The goals centered on a variety of emerging technologies, products, and services. But they also raised questions of how private stations might differ from the International Space Station. Would commercial stations prioritize profits over long-term scientific research? Would they focus on areas with immediate financial returns rather than projects with less obvious commercial applications? Advocates, however, point to the possibility for greater innovation, new business models, and faster adaptation than traditional models.

20 Apollo luminaries: If ever there existed a straight-to-Mars advocate, it was Dr. Robert Zubrin, who called for a more minimal "live-off-the-land approach" and who had co-drafted in 1991 a research paper that focused on "methane/oxygen bipropellant manufactured primarily out of indigenous resources." Again, talk of methane was in the mix. Methane power had been discussed at the highest levels since at least Rob had been in grade school, if not earlier. And Zubrin's paper was in part a response to President George H. W. Bush's Space Exploration Initiative. Announced in 1989, and overseen by Vice President Dan Quayle, the program was meant to develop a permanent Moon base while placing humans on Mars. But it seemed like an effort to redirect the energies and budgets of defense contractors, given the old Soviet threat had now vanished. A leaked ninety-day study put the price range up to $541 billion over four decades, and included an additional 55 percent "cushion" in expected cost overruns. That trillion-dollar

price tag would ultimately prove too much for Congress, and the idea lost the political momentum it needed. To learn more about Zubrin's "Mars Direct" plan, see the following: Robert M. Zubrin, David A. Baker, and Owen Gwynne, "Mars Direct: A Simple, Robust, and Cost Effective Architecture for the Space Exploration Initiative," Feb. 1991, doi:10.2514/6.1991-329.

20 Those like Senator Bill Nelson: Despite the shift, at least two facets of the second Bush legacy had remained: (1) the Orion spacecraft, used for Artemis; and (2) the bedrock for a privatized approach. In fact, in November 2005, under President George W. Bush, the NASA administrator, Mike Griffin, had created the Commercial Crew and Cargo Program Office. Rather than focusing on procurement contracts with principally legacy companies like Lockheed Martin, Northrop Grumman, and Boeing, the office was focused on partnering with companies investing their own capital. It then provided seed funding and technical expertise. Veterans of space, however, remained wary of shifting responsibilities to the private sector, particularly when it came to the perils of human spaceflight. Those like Marco Caporicci, head of the European Space Agency Transportation and Reentry Systems Division, called it unrealistic and warned of looming dangers to crewed flights. "You know what it means to lose a shuttle crew," he said. "No agency, no government, wants to go through that experience again."

20 Charles Miller: President Obama had even questioned the value of going to the Moon at all, as opposed to other celestial bodies, including Mars.

21 The industry seemed poised: Adding fuel to the fire in 2015, Congress passed the U.S. Commercial Space Launch Competitiveness Act, which effectively allowed private firms to explore and—in theory—mine asteroids for precious minerals and other space resources. President Obama signed the measure into law, with the intention of facilitating "a pro-growth environment for the developing commercial space industry." In an op-ed published by CNN, Obama explained how "just five years ago, U.S. companies were shut out of the global commercial launch market," and yet "today, thanks to groundwork laid by the men and women of NASA, they own more than a third of it."

21 For Steve, that was: The beginnings of private industry had, however, started much earlier. The founding of Blue Origin in 2000 and SpaceX in 2002, as well as Mojave Aerospace Ventures—led by the designer Burt Rutan—had given fresh legs to an industry long dominated by those legacy giants. In 2004, Rutan had made history when SpaceShip-

One, an experimental air-launched rocket-powered aircraft, became the first privately built crewed vehicle to reach the edge of space. Four years later, SpaceX's Falcon 1, a two-stage rocket capable of delivering a metric ton into low Earth orbit, became the cheapest such rocket in history. But it wouldn't be until 2015, when Blue Origin's New Shepard launched and landed vertically—and later that same year when SpaceX's Falcon 9 booster landed the first orbital-class reentry—that rocket reusability would truly begin to revolutionize the economics of spaceflight.

### CHAPTER 3: The View from China

22 when it came to the Moon: Chinese Academy of Sciences, "China Publishes World's First High-Definition Lunar Geologic Atlas," April 22, 2024.

22 that proposed facility: Andrew Jones, "China Plans to Build Moon Base at the Lunar South Pole by 2035," Space.com, Sept. 10, 2024.

22 "comprehensive lunar station network": Chinese efforts had also focused on small bioreactors that sprouted cottonseeds. In fact, the experiments had also yielded rapeseed and potato sprouts. Professor Liu Hanlong, who served as head of those experiments, later explained how potatoes could serve as the principal source of food for the first human settlement. Cotton could be used for clothing, while rapeseed could produce cooking oil. "We have given consideration to future survival in space," Liu told the *South China Morning Post*. "Learning about these plants' growth in a low-gravity environment would allow us to lay the foundation for our future establishment of [a] space base." Other, seemingly more radical ideas, like the growing of bacteria on carbon compounds found on asteroids to produce something that resembled "a caramel milkshake," were also being discussed.

23 By 2024, the nation: Denis Kalinin, "China: Private Space Ecosystem of the Rising Superpower," *Space Ambition*, Substack, April 25, 2025, spaceambition.substack.com.

26 Although, curiously, none of that seemed: That expectation was based on a partnership with ILOA and not Intuitive Machines.

26 Chinese observatory on the Moon: Ling Xin, "Chinese and US Scientists Build Bridges with Cutting-Edge Hale Telescope Project," *South China Morning Post*, July 25, 2022.

26 A decade later, with *Odie*'s launch: Similar collaborations also eventually occurred at California's Hale telescope, used to send and receive data encoded on a near-infrared laser some ten million miles away to

Psyche, a unique metal-rich asteroid between Mars and Jupiter. Chinese researchers were not involved in the Psyche mission, however.

27 Steve Durst, head of that Hawaii team: This was the result of a memorandum of understanding (MOU) signed between the National Astronomical Observatories of China (NAOC) and the Hawaii-based nonprofit International Lunar Observatory Association (ILOA) in 2012.

27 "He convinced me": Earlier, ILOA had signed a memorandum of understanding with China's Deep Space Exploration Laboratory to collaborate on China's budding lunar base, known as the International Lunar Research Station, which was considered a direct rival to the U.S.-led Artemis program.

### CHAPTER 4: Problems in Houston

30 the tardigrades were presumed dead: Flora Graham, "Tardigrades Didn't Survive Crash-Landing on the Moon," *Nature Briefing,* May 21, 2021.

30 "We knew there were risks": Daniel Oberhaus, "A Crashed Israeli Lunar Lander Spilled Tardigrades on the Moon," *Wired,* Aug. 5, 2019.

30 The Soviet's Luna 15 mission: Luna 15, NSSDCA/COSPAR ID: 1969-058A, NASA, Space Science Data Coordinated Archive.

30 *Odie*'s engine had to be sorted: "We're in a race," Steve told me one day in his factory, pointing to a massive American flag that hung over the din of machinists at work. "And we know they're watching us." By "they," he meant China. "We invest a lot to keep our data secure," he added, walking the factory floor. The company had initially contracted with the cybersecurity company IronNet, founded by the U.S. Army general Keith Alexander, who had run spy operations in Iraq and Afghanistan and later served as director of the National Security Agency. But that company had filed for bankruptcy just as reports of Chinese hackers burrowing into both America's critical infrastructure and private U.S. companies filled the headlines. A host of threats were accelerating out of eastern Europe, Iran, and North Korea. In 2022, Russia launched a cyberattack against Viasat, a satellite broadband services provider based in California, in a bid to undermine Ukrainian communications ahead of its invasion. Strikes can range from jamming and spoofing attacks, meant to mislead planes and ships with phony information, to electronic eavesdropping and denial of service assaults that disrupt service by overwhelming a system with a flood of traffic or data requests. And they can theoretically originate from almost anywhere, be it a state-sponsored operation or a lone hacker with a laptop. But of all America's adversaries,

it was Beijing's resources that were unrivaled. It was something Brandon Wales, then–executive director of the Department of Homeland Security's Cybersecurity and Infrastructure Security Agency (CISA), in 2023 had noted as a "significant change" when it came to Chinese cyber activity, which was focused on attacking not only infrastructure but also space infrastructure. In fact, in a two-page advisory, U.S. intelligence agencies warned that "foreign intelligence entities recognize the importance of the commercial space industry to the U.S. economy and national security, including the growing dependence of critical infrastructure on space-based assets." And yet curiously—in a series of moves years later under the second Trump administration—core parts of America's cyber defenses would soon be dismantled. Under pressure from a far-right political activist named Laura Loomer, General Timothy Haugh, who oversaw the NSA and U.S. Cyber Command, would soon be fired, alongside at least 130 other CISA personnel. Haugh had reportedly just submitted plans to scale up U.S. cyber-offensive efforts to Secretary of Defense Pete Hegseth, in part as a response to a 2021 Chinese-state-sponsored Volt Typhoon hack that had remained undetected in the U.S. electric grid for some three hundred days. A separate Chinese-sponsored attack, known as Salt Typhoon, was found to have broken into several major telecom companies, as part of an intelligence-gathering effort that targeted the U.S. 2024 elections. And yet the Cyber Safety Review Board, set up during the Biden administration to evaluate cyberattacks and provide recommendations, was almost immediately disbanded in President Trump's first month in office over what the acting DHS secretary, Benjamine Huffman, called the need to "eliminate misuse of resources." The board had not finished investigating the Salt Typhoon attacks prior to its dissolution.

### CHAPTER 5: A New Silk Road

34 cooperation on the peaceful use: Chinese Embassy Statement, "Hu Jintao Holds Talks with Argentine President," Nov. 17, 2004.

36 "We are interested in developing": "Russia, Argentina Sign Agreement on Space Cooperation," Tass, Oct. 8, 2019.

36 "It opened the door": It was the product of a bilateral agreement he helped engineer between the Chinese Moon Exploration Program, China's People's Liberation Army, and the Argentine space agency, CONAE.

37 BeiDou tracking stations: European Union's Galileo and Russia's Global Navigation Satellite System were also in development.

37 global positioning, navigation, and tracking: China's massive state-

owned companies, such as China Electronics Technology Group Corporation, were expected to further explore emerging markets, not to mention private Chinese tech companies, such as Huawei, Tencent, Alibaba, Baidu, Meituan, and Xiaomi, the latter of which had already invested heavily in the country's premier space companies (for example, GalaxySpace, ispace, and Deep Blue Aerospace). "Huawei Enters the Constellation Game?," *China Space Monitor,* Substack, Jan. 31, 2024, chinaspacemonitor.substack.com.

37 "embedding and serving": House Foreign Affairs Committee, "China's Objectives in Space, Regional Snapshot in Space," foreignaffairs.house .gov.

38 The Bayan Obo mines: Baochuan Li et al., "In-Situ Gamma-Ray Survey of Rare-Earth Tailings Dams—a Case Study in Baotou and Bayan Obo Districts, China," *Journal of Environmental Radioactivity* 151, pt. 1 (Jan. 2016): 304–10; Tim Maughan, "The Dystopian Lake Filled by the World's Tech Lust," BBC, April 2, 2015.

38 increase in demand by 2040: International Energy Agency, *The Role of Critical Minerals in Clean Energy Transitions,* World Energy Outlook Special Report, March 2022.

38 The so-called Lithium Triangle: Li et al., "In-Situ Gamma-Ray Survey of Rare-Earth Tailings Dams."

38 Local residents openly worried: Maughan, "Dystopian Lake Filled by the World's Tech Lust."

38 ablaze to blockade the entrance: Global China Unit, "Tensions Grow as China Ramps Up Global Mining for Green Tech," BBC, April 29, 2024.

38 Stanislav, for his part, was well aware: Stanislav also served as executive and technical director of the National Commission on Space Activities.

39 Now, however, with Roscosmos: There was also still collaboration on the ISS, even with Moscow's invasion of Ukraine in 2022, alongside U.S. accusations that Russia was developing nuclear-armed anti-satellite weapons in direct violation of the 1967 Outer Space Treaty. Moscow had denied the allegation, with President Vladimir Putin himself responding that "we have always been categorically against and are now against the deployment of nuclear weapons in space . . . [and] we urge not only compliance with all agreements that exist in this area, but also offered to strengthen this joint work many times." And somehow, Moscow and Washington continued to devise instances of doing precisely that. In 2024, for instance, despite ongoing Russian aggression in Ukraine and cyberattacks against the United States—and despite America's growing hand assisting Ukraine's fight against Russia—NASA and Roscosmos

agreed to continue launching astronauts and cosmonauts to the ISS. All indications were that it would continue until the station was crashed into the ocean by the end of 2030. Meanwhile, Russia continued to make headlines in space, both in success and in failure, including the crash of its Luna 25 spacecraft as it targeted its own lunar landing, and the first one thousand days in space achieved by the fifty-nine-year-old cosmonaut Oleg Kononenko, who by 2024 had orbited Earth sixteen thousand times and traveled some 420 million miles. Moscow clearly had a mixed record in space.

40 The Argentinians would provide: What Beijing really sought was a technology that would, among other things, make such ground stations obsolete, committing more than $15 billion in quantum computing investments that amounted to more than eight times what the United States publicly pledged to spend, as part of a space-based seemingly unhackable technology, the basis of which even Albert Einstein described as "spooky."

40 Javier Milei said that Argentina: "Argentina Presidential Frontrunner Calls China an 'Assassin,'" Bloomberg News, Aug. 17, 2023.

41 an official visit to Beijing: Igor Patrick, "China and Argentina in Early Talks over Javier Milei's Possible Visit to Beijing," *South China Morning Post,* June 15, 2024.

41 "He wouldn't travel to China": Patrick, "China and Argentina in Early Talks over Javier Milei's Possible Visit to Beijing."

## CHAPTER 6: The Path to Interplanetary Space

42 army rebels seized a nearby garrison: Argentina's "Dirty War" of the 1970s and early 1980s spawned more than three hundred secret prisons in a bid to detain political opponents and suspected "subversives." But especially in rural places like Las Lajas, which were far removed from the political wranglings of Buenos Aires, many officers felt that the pendulum had swung too far. In public statements that directly challenged the administration's central authority through an open rebellion, the coup essentially called for a "restoration of Army dignity." Three other garrisons joined their cause before the rebels were ultimately quashed.

42 Roving "death squads": Argentina Declassification Project: The "Dirty War" (1976–83).

42 declassified cables later revealed: "[Admiral] Massera sought opportunity to speak privately with me," the U.S. ambassador to Argentina, Robert Hill, said in a cable sent just a week before the 1976 overthrow of the Perón government, following a meeting with military officials.

"He said that it was no secret that military might have to step into political vacuum very soon."

42 As thousands "disappeared": Argentina's Military Coup of 1976: What the U.S. Knew, George Washington University Briefing book.

In February 1976, Assistant Secretary of State William Rogers reported to Secretary of State Henry Kissinger that "we would expect [the military government] to be friendly toward the United States." And yet "in stepping up the fight against the guerrillas, an Argentine military government would be almost certain to engage in human rights violations such as to engender international criticism."

43 "If there are things": Kissinger to Argentine Generals in 1976, National Security Archive declassified documents, George Washington University.

43 "That is, to terrorize": National Security Archive declassified documents, July 9, 1976, George Washington University.

43 "The quicker you succeed": Kissinger said this during a subsequent October conversation in New York, according to his memcon's verbatim transcript. "Kissinger to Argentines on Dirty War: 'The Quicker You Succeed the Better,'" National Security Archive declassified documents, George Washington University.

43 subsequent cables by Shlaudeman: "Kissinger to Argentines on Dirty War: 'The Quicker You Succeed the Better.'"

43 Mauricio Macri, would request: White House, Argentina Declassification Project, Dec. 12, 2016.

44 One of China's largest: According to interviews with local officials and hoteliers.

44 whose Chinese leadership was accused: Chatham House, "What Is China's Belt and Road Initiative (BRI)," Sept. 2021. The reality is more likely a bit more nuanced, with researchers like Johns Hopkins University's Deborah Brautigam and Harvard Business School's Meg Rithmire illustrating that Chinese banks have indeed shown a willingness to restructure the terms of existing loans, particularly when it comes to infrastructure projects in developing economies.

45 In truth, labor strikes: Simon Han and Jessica Song, "The Return of Strikes in China," *Asian Labour Review,* June 4, 2024.

45 The teams worked: Andrew Jones, "Report Highlights U.S. Concerns over China's Space Infrastructure in South America," *SpaceNews,* Oct. 6, 2022; "Un día en la estación espacial chino-argentina en Neuquén—Sorprendente Argenchina," CGTN, YouTube, 2000.

Chen Xue Jun, a senior engineer at the facility, meanwhile explained just how the state of Neuquén had met "the technical requirements for the electromagnetic environment," or quite frankly the sheer lack of electromagnetic interference found in the remote region.

45 other types of electromagnetic signals: Those civilian applications indeed appear to be real, such as the need for more effective tracking and monitoring of weather for a nation that is profoundly affected by weather disasters, be they storms, floods, or heat waves, which have historically cost the Chinese economy billions in damages and lost revenue. Other aspects of tracking, such as deforestation patterns, coastal erosion, traffic patterns to improve business functions, and even trash management, are indeed benefited by increased terrestrial connectivity to those circling constellations. The problem is, however, Chinese authorities are generally thought to have adopted surveillance tech, particularly in the aftermath of the COVID-19 pandemic, and have extensively collected both Chinese citizens' and foreign nationals' data, often with the goal of identifying and even predicting crimes and protests, as well as pressure points of adversaries and business leaders. High-bandwidth, real-time monitoring could indeed create a "nowhere to hide" paradigm, as Chinese scientists claim, with major implications in the realms of electronic warfare.

46 "well aware of": Natasha Bertrand, "Cuba Gives China Permission to Build Spying Facility on Island, US Intel Says," CNN, June 9, 2023.

46 Those concerns came to a head: Jeremy Herb et al., "US Officials Disclosed New Details About the Balloon's Capabilities. Here's What We Know," CNN, Sept. 2, 2023.

In response to concerns about the Cuba listening post, a Chinese Foreign Ministry spokesperson sought to turn the tables on growing criticism over China's surveillance efforts by referring to the United States as "the most powerful hacker empire in the world." In certain circles, that might be considered a not-so-subtle nod to the past efforts of a twenty-nine-year-old contractor named Edward Snowden, who stole and disseminated an enormous cache of highly classified documents, which revealed the sheer scope of America's surveillance efforts, including data collection on millions of Americans.

46 Intelligence gathering, be it: "Survey of Chinese Espionage in the United States Since 2000," CSIS, March 2023.

## CHAPTER 7: Artemis

47 As it meandered over: Jeff Foust, "Times Are Changing: NASA Looks to Move Beyond the Traditional Contract," *Space Review,* May 16, 2022.

48 SLS and the Orion capsule: Launch vehicles like SLS, in fact, had

grown so powerful that they had become a danger to themselves, producing vibrations so intense that they have at times been mistaken for earthquakes. SpaceX's Starship, when it was still in development, reportedly caused windows to crack up to five miles away from the launchpad. But the damage was not always farther afield. Such vibrations can damage the rocket itself. In fact, a NASA study from as early as 1971 determined that vibration and acoustic issues contributed to up to 60 percent of early rocket failures. To fix this, the agency would look to techniques from early stealth submarines devised around World War II, which employed air bubbles around the craft to absorb the acoustic waves of enemy radar. By shooting millions of gallons of water into the flame trench below the rocket, the effect worked to dampen the sound and vibrations inside.

49 In reality, the redesign: Mike Wall, "NASA's Huge New Rocket May Cost $500 Million per Launch," NBC News, Sept. 13, 2012.

49 "complexity of developing": NASA Office of Inspector General 2023 Report.

49 the system was even up: NASA's Management of Space Launch System Block 1B Development, NASA Office of Inspector General, Aug. 8, 2024.

50 Still, a full decade: Jeff Foust, "Parachute and Wiring Issues to Delay Starliner Crewed Test Flight," *SpaceNews,* June 1, 2023.

51 "deep financial exposure to China": Laura He, "Elon Musk Wins Official Praise for Tesla During Surprise Visit to China," CNN, April 29, 2024.

51 "successful domestically in China": Charles Lavery, "Taiwan 'Not for Sale': Elon Musk Rebuked for Toeing China's Line," *Newsweek,* Sept. 15, 2023.

### CHAPTER 8: Selling the Moon

53 "It is extraordinarily difficult": Gold, interview by Casey Dreier, Planetary Society, Sept. 2, 2022.

53 It was also "popular": That effort, known as the Asteroid Retrieval and Utilization Mission, was meant for a robotic spacecraft to travel to a near-Earth asteroid, grab a boulder from the asteroid's surface, and bring it to cislunar space, where astronauts could then conduct studies and collect samples. It was later canceled.

54 The 1967 Outer Space Treaty: China, like twenty-one other nations, has not ratified the treaty.

55 An analogy in U.S. Code: U.S. Code §51303: Asteroid resource and space resource rights.

**CHAPTER 9: Echoes of China's Rise**

57 In a way, they were wrong: Passed down throughout history, and present in as much as 20 percent of humans, the *DRD4*-7R gene variant has been linked with risk-taking behaviors and even the penchant to explore. Further studies link 7R and 2R to human migratory patterns, albeit controversially, in which one widely cited review, a first of its kind in 1999 carried out by Chuansheng Chen of the University of California, Irvine, examined eighteen indigenous populations along routes that delivered humans out of Africa and onto Europe, Asia, and across the Western Hemisphere. Chen found that the farther these groups traveled, the more probable it was that they carried these specific variants. The study has its critics, however, who suggest such linkages are overstated. And while it's certainly not known if any such markers ran through Zheng's veins, in the context of his value in Chinese history (and propaganda) as an irrepressible explorer, they might as well have.

57 "In the early 15th century": Ministry of Foreign Affairs, People's Republic of China, Speech by H.E. Xi Jinping President of the People's Republic of China at UNESCO Headquarters, March 28, 2014.

58 During that same speech: "Full Text of President Xi's Speech at Opening of Belt and Road Forum," Xinhua, May 14, 2017.

60 "What would the *Times*": Chris Bowlby, "The Palace of Shame That Makes China Angry," BBC, Feb. 2, 2015.

61 a common phrase in China: Andrew Moody, "Lessons of the Opium War," *China Daily,* Feb. 12, 2024.

61 China had made history: Huizhong Wu and Christopher Bodeen, "China Raises Defense Budget by 7.2% as It Pushes for Global Heft and Regional Tensions Continue," Associated Press, March 5, 2024.

61 Though the speech largely: Mike Wall, "China Moving at 'Breathtaking Speed' in Final Frontier, Space Force Says," Space.com, April 10, 2024.

62 Not only did it comprise: Frances Mao, "China's Far-Side Moon Mission Begins Journey Back," BBC, June 4, 2024.

62 "moon and other celestial bodies": G.A. Res. 2222 (XXI), Treaty on Principles Governing the Activities of States in the Exploration and Use of Outer Space, Including the Moon and Other Celestial Bodies, Jan. 1, 1967.

63 "under its jurisdiction": "Let's See China's Argument," China's Office of Policy Planning and Coordination on Territory and Sovereignty.

63 "I don't want them to get": Bill Nelson, "NASA's Administrator on Ambitions to Return to the Moon," interview with Scott Detrow, *All Things Considered,* NPR, May 5, 2024.

64 "The universe is an ocean": Manuel Moreno Minuto, "The Space Power of the Nations: A Maritime-Based Approach," U.S. Naval Institute, Dec. 10, 2020.

64 The small African nation had never signed: Fabio Tronchetti, "The 2019 Notice on Promoting the Systematic and Orderly Development of Commercial Carrier Rockets: The First Step Towards Regulating Private Space Activities in China," *Space Policy* 57 (Aug. 2021).

64 "Djibouti's lack of experience": Benjamin Silverstein, "China's Space Dream Is a Legal Nightmare," *Foreign Policy,* April 21, 2023.

65 Tiangong Kaiwu: Strategic mineral resources, placements of transport, and supply nodes include Earth-Moon Lagrange point 1, Sun-Earth L1 and L2, Ceres and Sun-Jupiter L1.

65 "Just like the miracles": Those comments were published in an official industry publication reported by China Space News on August 31, 2023; Chinese Society of Astronautics Report; Andrew Jones, "From a Farside First to Cislunar Dominance? China Appears to Want to Establish 'Space Economic Zone' Worth Trillions," *SpaceNews,* Feb. 15, 2020.

65 "At present, the United States": Andrew Jones, "Space Official Calls for China to Seize Crucial Opportunity to Establish Lunar Infrastructure," *SpaceNews,* March 31, 2023.

66 Fueling concerns was: "China Proposes Establishing Moon-Based Special Economic Zone," *China Briefing,* Nov. 8, 2019.

## CHAPTER 10: First Taikonaut

67 "struck [by] the technology": John M. Logsdon and James R. Millar, "U.S.-Russian Cooperation in Human Space Flight: Assessing the Impacts," Space Policy Institute and Institute for European, Russian, and Eurasian Studies, Elliott School of International Affairs, George Washington University, Feb. 2001.

67 "The Shenzou spacecraft shows": Relations between Russia and China would snag decades later with the war in Ukraine, particularly after President Vladimir Putin began making nuclear threats. Perhaps by then Russia's main space agency, Roscosmos, had lost its luster. By 2023, it had reduced its commercial satellite partnerships to a single craft sold to Angola. And yet, even though Beijing initially seemed to be backing away, Russian and Chinese strategic bombers continue to fly joint patrols over the western Pacific, while lunar and asteroid expeditions, as well as low Earth and deep space satellite infrastructure building, continued to lean on Russian know-how.

68 "You carry the dreams": John Pomfret, "China's First Space Traveler Returns a Hero," *Washington Post,* Oct. 15, 2003.

68 "I feel good": "China Sends Its First Astronaut, Yang Liwei, into Space on Board Shenzhou 5," *South China Morning Post,* Oct. 15, 2023.
68 "I felt my internal organs": Xinhua, "Yang Liwei Recall What It's Like in Space," *China Daily,* Sept. 25, 2008.
69 China offered comparatively bargain: Anatoly Zak, "Disaster at Xichang," *Smithsonian Magazine,* Feb. 2013.
69 Traveling out to evaluate: Zak, "Disaster at Xichang."
69 "I got out, turned": Zak, "Disaster at Xichang."
71 Heroes abound: "Statement from Apollo 11 Astronaut Michael Collins," *SpaceNews,* July 15, 2009.

### CHAPTER 11: Science in the Haunted 1950s

74 In 1990, attendees: History of Science Society Newsletter, Oct. 1990.
74 But on that day: Lawrence Badash, "Science and McCarthyism," *Minerva* 38, no. 1 (2000): 53–80.
75 "We've got to handle this": "The Scientist as Educator and Public Citizen: Linus Pauling and His Era," Transcript with Lawrence Badash, Oregon State University, Oct. 29–30, 2007.
75 "loathed McCarthy as much": John A. Farrell, *Richard Nixon: The Life* (Doubleday, 2017); Larry Tye, "The President and the Bully," *Saturday Evening Post,* Oct. 15, 2020.
76 "because of suspicion": "Passport Is Denied to Dr. Linus Pauling," *New York Times,* May 12, 1952.
77 "most consistently named": David Kaiser, "The Atomic Secret in Red Hands? American Suspicions of Theoretical Physicists During the Early Cold War," *Representations* 90, no. 1 (2005): 28–60.
78 Foreign-run cafés and theaters: Fortunately for him, he would leave before events turned violent. In fact, a three-month-long battle for the city would erupt just two years after Tsien left home, marking the outbreak of the Second Sino-Japanese War. The battle, waged by air, land, and sea, signaled the "totality of modern urban warfare," as the author Peter Harmsen put it in his 2019 accounting, titled "Armageddon Rehearsed: The Battle of Shanghai, August–November 1937." The fighting, he wrote, was "carried out with the entire array of modern weaponry, including tanks and aircraft, combined with a consistent disregard of the city's Chinese population, taken to extremes where civilians became all but invisible to the uniformed combatants." Subsequent Japanese atrocities during the Second Sino-Japanese War included mass executions, widespread rape, looting, and arson, particularly across the more inland city of Nanking, which was then the capital of Nationalist China. That history and intense rivalry, as well as gruesome human

experiments that would be conducted in Japanese-occupied Manchuria, would inform subsequent clashes over territory. They included the disputed East China Sea islands, which Chinese authorities say Japan had taken in 1895, as well as broader trade relations and indeed the competition in space, despite the normalization of ties in 1972. See Peter Harmsen, "Armageddon Rehearsed: The Battle of Shanghai, August–November 1937," in *A History of Modern Urban Operations,* ed. Gregory Fremont-Barnes (Palgrave Macmillan, 2020).

80 "How stark the contrast": Daniel Southerland, "The Genius Who Armed China," review of *Thread of the Silkworm,* by Iris Chang, *Washington Post,* Jan. 21, 1996; Milton Viorst, "The Bitter Tea of Dr. Tsien," *Esquire,* Sept. 1, 1967.

## CHAPTER 12: The Pentagon's New Ride

82 "That's not how rockets work": Mike Gruss, "Inside the Making of ULA's Next Rocket," *SpaceNews,* March 7, 2025.

83 "numerous German scientists": Declassified CIA report, Oct. 29, 2013.

83 "the guided missiles": Declassified CIA report, Oct. 29, 2013.

83 Soviet design bureau OKB-456: A. A. Siddiqi, "Rocket Engines from the Glushko Design Bureau—1946–2000," *Journal of the British Interplanetary Society* 54 (2001): 311–34.

83 Over time, like many Soviet-era: At the behest of those like the Soviet rocket engineer and program manager Valentin Petrovich Glushko, who in 1974 was named Soviet space program chief designer and who oversaw development of the Mir space station, the bureau would undergo a series of transformations in a bid to consolidate the Soviets' advanced propulsion efforts.

84 With its dual-nozzle engine: Brooke Mosley, "RD-180 Engine: An Established Record of Performance and Reliability on Atlas Launch Vehicles," 2011 Aerospace Conference, Big Sky, Mont.

84 "It's amazing that": Elon Musk (@elonmusk), Twitter, Feb. 10, 2019, 9:29 p.m.

84 By 1997, during a time: Mosley, "RD-180 Engine."

85 By then, the SpaceX Falcon 9: Those concerns, in fact, had come to a head in Ukraine in 2023 when SpaceX's founder, Elon Musk, refused to permit his company's Starlink internet service to be used to launch attacks against Russian troops in Crimea. The terminals were not a part of a military contract, given that Musk had been providing them free of charge to Ukraine as a result of the Russian invasion. But it underlined a broader vulnerability for the Pentagon of relying too much on a single commercial entity. "If we're going to rely upon commercial architec-

tures or commercial systems for operational use, then we have to have some assurances that they're going to be available," said the U.S. Air Force secretary, Frank Kendall, in the event's aftermath. "We have to have that. Otherwise they are a convenience and maybe an economy in peacetime, but they're not something we can rely upon in wartime."

85 Production delays, however: Most were national security related, although some were meant to support Amazon's growing Project Kuiper broadband constellation.

85 its cone was *Peregrine:* Jeff Foust, "Vulcan Centaur Launches Peregrine Lunar Lander on Inaugural Mission," *SpaceNews,* Jan. 8, 2024.

### CHAPTER 13: Moon Rush

86 "Machines talk to us": Danny Olivas post on LinkedIn.

87 Not the kind of thing: Human spaceflight naturally has different stakes altogether. Years earlier, on October 31, 2014, Virgin Galactic's reusable suborbital spacecraft—known as SpaceShipTwo—disintegrated and killed the test pilot Michael Alsbury, during what the company initially described as "a serious anomaly resulting in the loss of the vehicle." The craft had been carried to launch altitude by a quad-jet cargo aircraft before being released. The lift was meant to allow the rocket-powered craft to continue on to the upper atmosphere and beyond on the edge of space before returning to Earth. It was SpaceShipTwo's ninth test flight, and part of a space tourism and research venture founded by Sir Richard Branson, after he formed the company a decade earlier in 2004, helping to establish the mostly taxpayer-funded Spaceport America in Sierra County, New Mexico—billed as "the world's first purpose-built commercial spaceport," with the goal of eventually taking tourists and researchers on brief jaunts into space. The suborbital portion of the flight was meant to last only about two and a half hours. But as the ship reached supersonic speeds at an altitude of roughly fifty thousand feet, it broke apart, something the National Transportation Safety Board ultimately pinned on human error, which companies like SpaceX, Blue Origin, and Axiom Space, as well as Intuitive Machines and Astrobotic, would later try to mitigate by leaning more heavily on computing power and a series of automated maneuvers.

88 the rim of Shackleton crater: Shackleton Crater's Illuminated Rim & Shadowed Interior, NASA, April 20, 2022; V. T. Bickel et al., "Peering into Lunar Permanently Shadowed Regions with Deep Learning," *Nature Communications,* Sept. 23, 2021.

89 "The United States of America": Ian Sample, "US 'Ready to Fight in Space if We Have To,' Says Military Official," *Guardian,* May 28, 2023.

89 A journal article: Zhang He et al., "Proposals for Sites Selection of Soft Landing on Lunar South Polar Region," *Journal of Deep Space Exploration* 7, no. 3 (2020): 232–40.

In coming years, Chinese authorities would establish near Shanghai what they referred to as China's Deep Space Exploration Laboratory, detailing a far more expansive road map to explore, establish settlements, and conduct scientific research across the solar system. The plan included a 2029 mission to explore Jupiter and its moon, Callisto; a 2033 mission to Venus, meant to return atmospheric samples similar to an upcoming Rocket Lab–MIT mission; a 2038 research station on Mars; and a 2039 nuclear-powered mission to Neptune's largest moon, Triton. But when it came to China's national space program, many of those efforts began with the Moon, which could theoretically be used to harvest water for fuel and build the sort of infrastructure that might accelerate those broader plans. China's lunar program had effectively started in 2007 with the Chang'e 1 lunar orbiter. Since then, the country's advancing space and military expansion led to a 2018 U.S. National Defense Strategy that labeled it a "strategic competitor." Given that, and the advancing timetable, the NASA administrator Nelson was now urging Steve to pick up the pace.

89 "Can you get there": This reporting relies on Steve Altemus's recollection of events. Bill Nelson could not be reached to independently confirm his statements, and this reported exchange should therefore be considered in the context of information derived from a single source.

89 Two of its modules: The Soviet Union's Luna 24 robot performed a similar mission in 1976.

90 The Chang'e 6 mission: Andrew Jones, "Chang'e-6 Delivers First Lunar Far Side Samples to Earth After 53-Day Mission," *SpaceNews,* June 25, 2024.

90 The priority then seemed to focus: Jeff Foust, "Northrop Grumman Takes $36 Million Charge on NASA Gateway Module," *SpaceNews,* July 27, 2023.

90 Then, in July 2024: Of course, even that was questionable. In 1984, for instance, before the U.S. Space Station Freedom project evolved into the International Space Station, the eventual $100 billion price tag had been listed at a mere $8 billion. Mission and cost creep were real concerns, as were delays. The project was initially meant to launch in 2022, a date that had since been pushed to at least 2027, and soon after employed in the Artemis 4 mission—the program's second attempted Moon landing.

90 "several significant challenges": "Artemis Programs—NASA Should

Document and Communicate Plans to Address Gateway's Mass Risk," Report to Congressional Committees, U.S. Government Accountability Office, July 31, 2024.

### CHAPTER 14: Red Star and Clover

92 Eight thousand miles west: In reality, movement of the Sun across the sky is a phenomenon based on perspective. Earth spins on its axis from west to east and thus creates the illusion of our star moving across the sky, a fact certainly not lost on many of those descending on this commercial heart of central China that was now making a bigger push into space.

92 commemorated the launch: In 2016, China had designated April 24 as its official Space Day.

92 Mars and Jupiter: "Wuhan Aims to Become China's 'Valley of Satellites' in Space Initiative," Reuters, March 17, 2022.

93 By the summer of 2023: Ling Xin, "Chinese Scientists Hold First Conference to Discuss Building Crewed Moon Base," *South China Morning Post,* April 12, 2023.

94 such punishing environments: Xin, "Chinese Scientists Hold First Conference to Discuss Building Crewed Moon Base."

94 "Humans evolved living": DIVINER Lunar Radiometer Experiment, UCLA.

95 "Now is the critical time": Jones, "Space Official Calls for China to Seize Crucial Opportunity to Establish Lunar Infrastructure."

### CHAPTER 15: *Peregrine*

96 this four-legged, foil-encased: The cremated remains and DNA of more than seventy people, including the science fiction writer Arthur C. Clarke, who cowrote the screenplay for the 1968 film *2001: A Space Odyssey,* were also on board.

96 "We know we're headed": Foust, "Vulcan Centaur Launches Peregrine Lunar Lander on Inaugural Mission."

97 managed to get the craft: Eric Berger, "The Situation with Astrobotic's Lunar Lander Appears to Be Quite Dire," *Ars Technica,* Jan. 8, 2024.

97 "Unfortunately, it appears the failure": UPDATE #4 FOR PEREGRINE MISSION ONE, Astrobotic, Jan. 8, 2024.

97 private lunar lander had failed: Between 1969 and 1972, twelve Americans had walked on the Moon. In fact, "people were getting bored," recalled Jim Lovell, who served as Apollo 13's command module pilot. Half a billion people had tuned in to the Apollo 11 Moon landing. But by NASA's third attempted landing, live transmissions beamed down

from Apollo 13 were not even picked up by major television networks. "The publicity for Apollo 13 you could find on the weather page of the newspaper, that was it," Lovell said. (That, of course, changed when an oxygen tank in the service module exploded and threatened both the mission and the lives of those on board.)

98 In just the past few years: For those that did successfully make it to the surface, landers would have to descend with almost no atmosphere to slow them down, blinding amounts of dust kicked up by the craft's thrusters, and no global positioning systems for guidance.

### CHAPTER 16: Final Descent—*Columbia*

99 Two thousand three: NASA, "20 Years Ago: Remembering Columbia and Her Crew," Jan. 26, 2023.

99 "at least five 'escapes' ": Mary Pat Flaherty et al., "Columbia Marked by Delay, Faults," *Washington Post,* Feb. 23, 2003.

99 The veracity of that reporting: David Rose, "How Warning Signs Were Ignored Before Disaster Shuttle's Launch," *Guardian,* June 21, 2003.

101 Frenchies is an Italian restaurant: Nick Natario, "Astronauts Say They Wish They Had This Restaurant's Food in Space," ABC13, April 17, 2020.

102 "The issue is not even": Ralph Vartabedian, "E-Mail to Columbia Discounted Danger," *Los Angeles Times,* July 1, 2003.

104 The subsequent accident report: Columbia Accident Investigation Board, Report, vol. 1, Aug. 2003.

104 "We cannot find any": "Bush Cancels Space Shuttle Program," *Space Daily,* April 1, 2005.

104 "incredibly expensive and really": Tim Fernholz, "How the Columbia Tragedy Began the Age of Private Space Travel," *Quartz,* Feb. 1, 2018.

### CHAPTER 17: Moon Mountain

107 More than 16,400 feet tall: There are other factors under consideration for development of a permanent human presence on the lunar surface, including the effects of solar radiation on human health. The Moon generally does not emit a magnetic field to protect it from charged particles. And yet there are indeed isolated lunar locales that *do* emit such magnetic fields, which create small pockets of protection that could work to shield future human settlements. Those fields also tend to coincide with the lunar swirls that are often visible with backyard telescopes. Researchers have conflicting hypotheses as to why those fields exist at all, often discussing the possibilities of asteroid impacts that bring magnetic anomalies to the lunar surface. But there is grow-

ing evidence that points to the streams of cooled magma beneath the lunar surface, which suggest lunar volcanic activity billions of years ago. Growing attention has focused on the content of those streams, which may include a titanium-iron oxide mineral called ilmenite, which under certain conditions can emit a magnetic field. If strong enough, those fields could protect the lunar surface from the kind of solar radiation considered hazardous to long-term human settlements. Ultraviolet images captured by the Lunar Reconnaissance Orbiter Camera found two large groupings of lunar swirls in the Marginis-King and South Pole–Aitken Basin, the latter of which is where Chinese exploration has focused since its early Chang'e missions.

107 Yet if they could get there: Shackleton crater was named after the British explorer Sir Ernest Henry Shackleton, who led Antarctic expeditions and endeavored to reach Earth's South Pole.

107 We had spoken briefly about it by phone: In 2007, Durst had become director of the very International Lunar Observatory Association responsible for the pair of cameras aboard *Odie,* whose images were intended to be shared with Chinese astronomers. China's first lunar lander, Chang'e 3, had snapped images of a pinwheel galaxy using its Lunar Ultraviolet Telescope. And the Apollo 16 lunar module pilot Charlie Duke and commander John Young had used a Far Ultraviolet Camera and Spectrograph to gaze at star clusters and distant nebulae. But that sort of thing had never been done in a truly permanent way.

108 A. P. J. Abdul Kalam: John F. Burns, "Self-Made Bomb Maker: Avul Pakir Jainulabdeen Abdul Kalam," *New York Times,* May 20, 1998.

108 And if that could be done: Of course, India was also not immune to questions about space spending amid domestic problems on Earth. In a video that went viral years later, after the Chandrayaan-3 touched down near the lunar south pole, a man quipped that "we are already living on the moon." When the interviewer asked how, he responded, "Does the moon have water? No. And we also don't have water. Does the moon have gas? No. Neither do we." It turns out that NASA's SOFIA mission confirmed lunar water does in fact exist and that it could also be converted into "gas" or fuel. It also turns out that the man might have actually been Pakistani and referencing the crescent Moon on his nation's flag. But the comments nonetheless pointed to a very real policy debate emerging over allocation of the nation's finite resources. It also reminded some of the early backlash over NASA's Apollo lunar program, a $283 billion effort in inflation-adjusted dollars, particularly from those like Gil Scott-Heron and his 1970 poem titled "Whitey on

the Moon." "I can't pay no doctor bills," the poem read, "but whitey's on the Moon. Ten years from now I'll be payin' still, while whitey's on the Moon."

### CHAPTER 18: The Makings of China's Commercial Rockets

110 "had not seen the true": Dan Schawbel, "Neil deGrasse Tyson: How He Became America's Top Astrophysicist," *Forbes,* May 19, 2017.

110 in a similar way, Changsha: In 2023, the city commanded attention when a U.S. sanctions list was updated to include a local satellite manufacturer, which had allegedly supplied a Russian paramilitary organization, known as the Wagner Group, with radar satellite imagery of Ukraine. But somehow the headlines seemed to understate the growing influence of the city itself, with its increased emphasis on manufacturing and electronics, particularly materials sciences and advanced computing.

110 Once governed by the rhythms: "Hunan Leaps Forward in Poverty Fight," State Council Information Office, People's Republic of China, Aug. 6, 2022.

110 having spanned some three thousand: The Silk Road was a series of Eurasian and North African trade routes, which began long before the birth of Christ, and was named as such because of the increased value and export of Chinese silk along these routes. Silk, however, was only part of the trade. Spices, precious metals, and minerals, as well as cultural performances involving dance and music, and intellectual exchanges, played significant roles in the global dissemination of culture, science, philosophy, and religion. Current Chinese leadership has used the concept of a "space" silk road to mirror those ancient exchanges in a more modern context, applying the principles of connectivity and exchange to the realms of commercial space development, exploration, and satellite communications.

111 While it's not clear: Lanxin Xiang, *Mao's Generals: Chen Yi and the New Fourth Army* (University Press of America, 1998).

111 as China's rapid military: Between 2000 and 2016, China's military budget increased annually at roughly 10 percent, according to a 2018 U.S. Defense Intelligence Agency report, with a defense budget of approximately $230 billion in 2022—eclipsed only by the United States. And yet those numbers may betray the extent of the difference. In 2022, the International Monetary Fund determined China's purchasing power parity was nearly a full quarter (23 percent) larger than that of the United States, given China's lower costs of goods and

services, exchange rate fluctuations, government policies, and large manufacturing base. That means Chinese currency essentially stretches further domestically than its American counterpart. That, combined with the relative opaqueness of actual Chinese military spending, may indeed close the gap when it comes to U.S.-China military spending. In 2025, the research firm Euroconsult estimated that Chinese government space spending had swelled to $20 billion in 2024.

111 a natural fit: Ministry of National Defense of the People's Republic of China, "PLA Embraces a New System of Services and Arms," April 19, 2024.

112 quantization, which essentially constrains: Quantization is essentially the process of mapping infinite values to smaller sets of finite values, whereas "weight pruning" is an optimization technique that shaves off, and effectively sets to zero, the least valued weights in the network. Finally, Huffman coding is essentially a data compression algorithm that keeps the integrity of data, even as its relative size is reduced.

## CHAPTER 19: Imagining Tomorrow

114 Between 1959 and 1961: Vaclav Smil, "China's Great Famine: 40 Years Later," *BMJ* 319, no. 7225 (1999): 1619–21.

115 as a three-body problem: Centuries earlier, Sir Isaac Newton earned his acclaim for solving the so-called two-body problem by building on the work of the German astronomer and mathematician Johannes Kepler, explaining how gravitational attraction locks planets into stable orbits. Albert Einstein's theory of general relativity explains how mass and energy bend space-time, and how those interactions effectively predict the path of objects like suns, planets, and moons under the influence of gravity. Those dynamics are illustrated not only in our own solar system but also in binary star systems, where pairs of stars revolve around a common center, and can effectively keep planets in their collective orbit. The first such system, in fact, was discovered in 2020, drawing comparisons to the fictional *Star Wars* planet of Tatooine, where the Luke Skywalker character stares out toward a binary sunset. See Jeanette Kazmierczak, "NASA's TESS Mission Uncovers Its 1st World with Two Stars," NASA, Jan. 7, 2020.

115 Rooted in scientific principles: Our closest neighboring star system, Alpha Centauri, is actually a stable triple-star system. Two of those stars—Alpha Cen A and Alpha Cen B—are in fact near twins of our own Sun and relatively close to each other. However, that third star—Alpha Cen C (often known as Proxima Centauri)—is a much smaller

red dwarf that has an orbit NASA estimates to be "more than 10 thousand times farther from the AB pair than the Earth-Sun distance." While it may seem as if this undermines the veracity of the three-body problem, it is assumed workable because Proxima is so much smaller, and orbits so far away from Alpha A and B, and that its gravitational influence is relatively weak and thus allows for a comparatively stable three-star system. There are also other questions concerning the novel's premise, such as how it's not clear how long planets in an unstable three-star system would survive before being destroyed or thrown out into space. And, for the purpose of the novel, that raises questions as to whether there'd actually be enough time to develop an advanced alien civilization, given the billions of years it took life on Earth to develop from simple organisms to more complex creatures to advanced technological societies. Still, it makes for a compelling read.

115 a story homegrown in China: Described by Einstein as "spooky action at a distance," the concept of quantum entanglements explains how two or more particles can exist in a so-called entangled state, whereby certain aspects of one particle rely on and affect another, no matter the distance between them. Entanglements in the quantum state effectively upend our most basic perceptions of reality, at least at a particle level. Meanwhile, the idea of a nonlocal universe—meaning not only that objects are affected by what's in their surroundings but that that measurement itself may influence an object's properties—was growing. In a study published in the scientific journal *Nature,* researchers concluded over the course of a series of controlled experiments using tiny electron detectors that just the extent of the devices' observation prompted changes in the patterns of the electron waves. It's a phenomenon known as the observer effect—the notion that particles can behave differently based on whether or not they are being watched. Such notions are unsettling, and cast uncertainties about what we assume and take for granted about the world in which we live, even to Einstein himself. "Do you really believe the moon is not there when you are not looking at it?" he reportedly quipped to a friend.

115 And for Song, when he read it, the effects: In fact, researchers have been exploring ways to devise architecture for ultrafast long-distance quantum communication that leverages entanglements. In September 2024, Boeing had even announced plans to demonstrate quantum entanglement swapping for the purpose of a communications network that connects quantum computers and sensors, which showed groundbreaking potential, having apparently already demonstrated the tech-

nology during tabletop experiments at research facilities in Malibu, California. "An array of telescopes around the Earth, connected with a quantum network, could go and image one of the rovers on Mars," Makan Mohageg, the principal applied quantum physicist at Boeing, told *Aviation Week*. "It's really wild."

115 "I think *Three-Body Problem*": In quantum computing, qubits can exist in what's called a superposition, which effectively means that they can represent both 0s and 1s at the same time. Much unlike a classical bit, which is a basic unit of data, qubits when grouped together represent a quantum system that could perform parallel computations that yield far greater computational power than classical computing. Nanophotonic diamond cavities, meanwhile, are essentially just those nano-scaled structures that contain certain properties, which could allow scientists to confine and manipulate photons for use in quantum information processing.

115 "I was born in 1992": A Netflix adaptation prompted mixed responses in China, with some claiming the new series strayed too far from Liu's original work.

117 the Internet of Things: "What Is the Internet of Things (IoT)?," McKinsey & Company, May 28, 2024; Ivica Prlić et al., "Wi-Fi Technology and Human Health Impact: A Brief Review of Current Knowledge," *Arh Hig Rada Toksikol* 73, no. 2 (2022): 94–106.

119 "Musk has more ambitious goals": Andrew Jackson, "Why This Veteran Neuroscientist Is Skeptical of Elon Musk's Neuralink," *Mars,* Sept. 11, 2020.

119 a thirty-year-old quadriplegic: Note that Arbaugh is not the first person to receive a BCI. Nathan Copeland had received one nearly a decade earlier, although the device was significantly different from the Neuralink device.

119 If Arbaugh merely thought: There are some concerns, derived from separate interviews within the neuro-research community, about Neuralink's degree of transparency.

120 a remarkable pace of advancement: Rachael Levy, "Neuralink Implanted Second Trial Patient with Brain Chip, Musk Says," Reuters, Aug. 5, 2024.

120 "researchers to run experiments": Fred D. Jordan et al., "Open and Remotely Accessible Neuroplatform for Research in Wetware Computing," *Frontiers in Artificial Intelligence,* May 2, 2024.

120 In the context of space: "Wetware computing and organoid intelligence is an emerging research field at the intersection of electrophysiology and artificial intelligence," the paper noted. "The core concept involves

using living neurons to perform computations, similar to how Artificial Neural Networks (ANNs) are used today. However, unlike ANNs, where updating digital tensors (weights) can instantly modify network responses, entirely new methods must be developed for neural networks using biological neurons." Meanwhile, Neuralink, in order to function, had developed flexible "threads" that could effectively be sewn into a human brain in a bid to initially help those suffering from paralysis but to one day, as Musk described it, "achieve a symbiosis with artificial intelligence."

The ambition was to effectively assist humans in "merg[ing] with AI" so that humanity wouldn't be "left behind" as AI grew more advanced. Like many aspects of Musk's endeavors, there was also a bent toward space. In the peer-reviewed quarterly journal *Neuroscience Research Notes,* researchers examined how BCI could one day assist astronauts with complicated tasks, communications, and discovery while improving health and well-being. See "An Era of Brain-Interface: BCI Migration into Space," *Neuroscience Research Notes,* June 2, 2021.

120 "achieve a symbiosis": Sigal Samuel, "Elon Musk Reveals His Plan to Link Your Brain to Your Smartphone," *Vox,* July 17, 2019.

120 And Song: Tamara Hummadi and Indranath Chatterjee, "An Era of Brain-Computer Interface: BCI Migration into Space," *Neuroscience Research Notes,* June 2, 2021.

120 Musk was the model: Jordan Kinard, "These Living Computers Are Made from Human Neurons," *Scientific American,* Aug. 8, 2024.

121 "What is consciousness?": In 2023, his company partnered with Matrix AI Network, a leading firm in decentralized AI investments, where Song had briefly worked, with the idea of utilizing the company's AI-powered blockchain with NeuraMatrix's implant technology to develop new concepts, using a chip system similar to those developed at DeePhi. "We needed low-power conception chips and with low temperature, so that can be planted into your brain [without health effects]," Song explained. "[Too high of a temperature] and it will burn your brain."

The goal, like at Neuralink, was to create a wireless channel of data exchanges between the brain and a technology that the human body wouldn't reject. But, like Musk, Song was ready to make the switch to rockets, co-found a launch company of his own, and devise a way to reach space semi-independently.

## CHAPTER 20: The Night Before Blastoff

122 Tired, and flanked by two: Steve had insisted on flying out of launchpad 39A, which SpaceX leased from NASA for its Falcon 9 and Falcon

Heavy launches. Apollo missions had all launched out of 39A. Starting in 1979, every space shuttle mission had also used the facility. That is, until 1986, when Space Shuttle *Challenger* rolled out of launchpad 39B with tragic results. The failure of an O-ring in one of the rocket boosters prompted the vehicle to break apart just seventy-three seconds after launch, killing all seven crew members on board, including Christa McAuliffe—a schoolteacher meant to be the first private citizen in space. And yet today, 39A couldn't be used as is. Given *Odie*'s novel liquid methane and oxygen propellant engine, the pad would have to be modified to permit fueling while the vehicle sat inside the fairing—a first for SpaceX—given its cryogenic fuels could be stored for only a limited time prior to launch.

125 the craft employed linerless tanks: Usually, the lining in the tank was employed to protect its walls from being corroded from the inside, which could lead to cracks. It was a propellant leak that was believed to have doomed Astrobotic's *Peregrine* vehicle. Liners can be weighty, however. And in the consummate trade-off between weight and performance, IM-1 had discarded the liners in favor of using a relatively new composite, apparently capable of withstanding corrosion and the pressure buildup from the methane propellant cooking inside. "Every kilogram that we can save," explained Rob, "is a kilogram we can sell."

125 continuous thrust propulsion: Traditionally, the craft's alignment during lunar descent is autonomously adjusted by way of short-duration pulses—usually thought of as the most precarious phase of the journey. Even vehicles exhibiting few problems have still ended up crashing into the lunar surface. In August 2023, when Russia's Luna 25 lander, the country's first such vehicle since the old Soviet days, entered lunar orbit, the craft suffered what Roscosmos described as an "emergency condition" after a malfunction in the firing of the engine, causing the craft to have "ceased its existence as a result of a collision with the lunar surface."

India's Chandrayaan-2 mission had met a similar fate during what the former Indian Space Research Organization (ISRO) chairman K. Sivan dubbed the craft's last "15 minutes of terror." The vehicle appeared to deviate from its trajectory, and through a series of malfunctions could not control its speed as it plunged toward the Moon. "The craft had to make very fast turns. When it started to turn very fast, its ability to turn was limited by the software because we never expected such high rates to come," said the ISRO chief, S. Somanath. As the vehicle scoured the lunar surface for its landing site, the craft was "trying to reach there by increasing the velocity," he added. Even though it was close to the ground, it "kept on increasing the velocity" and ultimately crashed.

IM-1 was using a different approach. Instead of pulsing its way down, which would require repeatedly firing engines as it descended, or simply shutting off certain engines, *Odie* would maintain a continuous thrust on descent.

126 The craft relied on: Mars Science Laboratory: Curiosity Rover, NASA, science.nasa.gov.

The system was meant to provide *Odie* with ample flexibility when it came to scheduling. The sheer increase in space missions had left NASA's Deep Space Network "extremely busy trying to track all of these space missions at once," according to the agency, which also provided Intuitive Machines with a capacity to sell commercial services to other companies and even agencies for their respective space missions. The idea was to create a sustainable company not wholly reliant on NASA's prerogatives.

127 "was a possibility of unauthorized": "Japan Space Agency Hit with Cyberattack, Rocket and Satellite Info Not Accessed," Reuters, Nov. 28, 2023.

127 "Elon Musk, we will": Dark Web Intelligence, X, March 14, 2023, 10:00 a.m.

127 By August, the FBI: Director of National Intelligence, "Safeguarding the US Space Industry: Keeping Your Intellectual Property in Orbit," National Counterintelligence and Security Center.

128 106,000-square-foot operational hub: Gone were the days of testing rockets in the open beside Ellington Airport. The new "flame range" consisted of a thirty-eight-hundred-square-foot reinforced-concrete chamber that was surrounded by a twenty-five-foot-high perimeter wall. Having it on-site and next to manufacturing operations in theory allowed engineers to make quick adjustments and incremental improvements to the vehicles. Speed and efficiency were baked into the design. But so was security.

### CHAPTER 21: Launch from the East

129 Standing at roughly thirty meters: Or 3.7 tons to Sun-synchronous orbit at an altitude of seven hundred kilometers.

130 the sheer scale of its Starlink constellation: Starlink had plans for as many as forty-two thousand satellites, while Amazon's Project Kuiper had launched thousands of its own satellites for a broadband megaconstellation that it hoped would rival Starlink.

130 "prevent the Starlink constellation": Stephen Chen, "China Aims to Launch Nearly 13,000 Satellites to 'Suppress' Elon Musk's Starlink, Researchers Say," *South China Morning Post,* Feb. 24, 2023.

130 "For China to establish": Ling Xin, "China Boosts Its Project GW Satellite Rival to Compete with Elon Musk's Starlink," *South China Morning Post,* July 31, 2023.

131 leadership now seemed to recognize: Chinese private sector companies are typically defined as those with less than 10 percent state ownership.

131 To succeed, particularly on the military: In a sign of just how fast Beijing was now moving, within a year of NASA's SLS launch Chinese scientists had already touted the actual construction of a "revolutionary" air-breathing engine, which could potentially deliver a craft into the stratosphere at roughly sixteen times the speed of sound. That was not quite the speed needed to reach orbit. But it could, the researchers noted, "improve the optimal thermodynamic cycle efficiency in nearly all speed ranges" while bringing about "a revolutionary change in aerospace propulsion." China, meanwhile, would soon unveil plans to take hypersonic spaceflight a step further. The design, devised by CASIC's Flight Vehicle Technology Research Institute, called for the combining of hypersonic air-breathing technology with an electronic launch track, which could potentially accelerate a spacecraft to Mach 1.6 before the vehicle itself detached from the track and into space. As it ascended, the craft's engines would fire, increasing its speeds and taking the vehicle the rest of the way. The system was essentially devised to supplant the usual need for massive quantities of fuel used in traditional rocketry, potentially eliminating the need for a first-stage rocket, and promised—if successful—to dramatically reduce future launch costs. To test it, CASIC had already constructed a more-than-one-mile track in Datong, Shanxi province, effectively borrowing a derivative of the technology employed in the nation's high-speed train systems, to fling objects skyward at more than six hundred miles per hour. The expectations were for longer tracks, larger objects, and higher speeds. Ironically, it was NASA, more than three decades earlier, that devised the method, and first attempted it, albeit with a much smaller (less than fifty-foot) track. That system, however, for want of funding, ultimately led to a scrapping of the project.

## CHAPTER 23: Lunar Payload

140 "risk of a catastrophic": Carson Ezell, Alexandre Lazarian, and Abraham Loeb, "A Lunar Backup Record of Humanity," arXiv, Sept. 22, 2022.

140 To do that effectively: Others weren't so sure. Never mind the high cost

and logistics involved, as well as limited infrastructure, latency, and accessibility. "If I had a huge amount of data and I didn't want anybody to ever see it, there are bunkers," Matt Konda, CEO of the cybersecurity firm Jemurai, explained over the phone. "Why would it need to be on the Moon? Distance, in the hacker sense, didn't quite matter." "If it's connected," he added, "then it's exposed."

140 "It might take some time": Mars Science Laboratory: Curiosity Rover, NASA.

140 Ramon.Space: Ramon.Space, alongside Lulav Space, a robotics company that specialized in navigation sensors for space applications, had collaborated for Beresheet 2, a private space mission meant to land two spacecraft on the Moon in 2025, with two landers deployed from an orbiter. And yet, in 2023, the funding for those missions, which had largely leaned on philanthropic support, seemed to be fading. "These times obligate us to invest our resources and time in other philanthropic projects," said the billionaire Morris Kahn, who funded the first failed Beresheet mission. The fate of the second mission was at least initially unclear as other agencies and companies raced to plant their craft in the lunar soil.

140 the IM-1 mission was a proof: It was essentially a software trial run in anticipation of a second lunar IM mission, when Lonestar hoped to deliver a full Yocto Linux system, with 8 terabytes of storage, to the Moon's surface. "This all came from customers," Chris told me. "We're not a technology-push company. This is a demand pull."

141 Considered some of the most: Data Centers and Servers, U.S. Department of Energy, FAQ.

141 Growing consumer demands: Matthew Gooding, "Newmark: US Data Center Power Consumption to Double by 2030," Data Center Dynamics, Jan. 15, 2024.

142 "It is as significant": Zhang Tongi, "China Plans to Build 'Three Gorges Dam in Space' to Harness Solar Power," *South China Morning Post,* Jan. 9, 2025.

142 Russia's space program experienced that: Louis D. Friedman, "Phobos-Grunt Failure Report Released," Planetary Society, Feb. 6, 2012.

142 The problem was that some: Laura Pernigoni, Ugo Lafont, and Antonio Mattia Grande, "Self-Healing Materials for Space Applications: Overview of Present Development and Major Limitations," *CEAS Space Journal* 13 (2021): 341–52; Rachel Rose, "Designing Self-Healing Chips to Withstand Space Radiation," Texas A&M University, June 8, 2022.

143 a lander, hopper, orbiter: A joint venture between Intuitive Machines

and the Maryland-based nuclear reactor developer X-energy had received $5 million in Department of Energy and NASA contracts to design a fission power grid that would deliver at least 40 kWe of power to the Moon by 2028, about the amount of energy needed to simultaneously power eighty Tesla Model S vehicles. The system, again if successful, could prove especially useful in providing electricity to habitats, mining operations, and rovers, as a part of plans for a prospective Artemis base camp. By December 2023, the company had raised $235 million in a Series C funding. Kam Ghaffarian would be among those leading such efforts. A year later, Amazon announced a $500 million funding round for X-energy, focused on domestic power generation over the next two decades. But the application for space was also clear.

143 an explosion of data: China's existing domestic system is sweeping, averaging one camera per every two citizens.

143 systems might be devised: Stephen Chenin, "Skynet 2.0: China Plans to Bring Largest Surveillance Camera Network on Earth to the Moon to Protect Lunar Assets," *South China Morning Post,* March 4, 2024.

143 "We can't beat communists": Steve Kwast, "How American Entrepreneurs Can Help Win the Space Race Against China," *SpaceNews,* Dec. 4, 2024.

143 Dennis Wingo was the founder: Kevin Krewell, "RISC-V Summit Expands Influence, Shows Growing Pains," *EE Times,* Nov. 20, 2023.

## CHAPTER 24: Huntsville

147 A 1989 interview: Stuhlinger, interview by A. Dunar and S. Waring, MFSC History Project, NASA, April 24, 1989.

147 From there, Stuhlinger: In a speech he delivered at the Army Science Conference at West Point in July 1957, Stuhlinger noted that the three-hundred-pound payload of the Jupiter-C reentry test vehicle missile, a special modification of the Redstone ballistic missile that had been designed and launched by the Army Ballistic Missile Agency, could be converted to a fourth rocket stage that could then carry an artificial satellite into orbit.

147 But in those early days: The U.S. Army's Redstone Arsenal, which served as a hub of American missile programs, was the birthplace of the Redstone rocket, used as part of NATO's Cold War shield of Western Europe, which also carried Alan Shepard as the first U.S. astronaut into space.

147 These fields would have to be: "Nuclear Propulsion Could Help Get Humans to Mars Faster," NASA, Feb. 12, 2021.

147 more compact and complex circuitry: Jaime Lightfoot, "Introducing

ENIAC Six: Atomic's Room Named for the Women Who Programmed the ENIAC," Atomic Object, July 31, 2016.

147 Used to amplify or switch: That approach had in 1945 powered the world's first high-speed computer, a gargantuan piece of machinery built by the University of Pennsylvania used to calculate ballistic trajectories. But it weighed thirty tons, occupied a fifteen-hundred-square-foot highly air-conditioned room, and housed more than seventeen thousand vacuum tubes, each of which burned out at a rate of roughly one per day. Though the army proclaimed that ENIAC (the Electronic Numerical Integrator and Computer) was "expected to revolutionize the mathematics of engineering and change many of our industrial design methods."

148 The technology offered a range: Margaret Hamilton, the pioneering computational scientist who led the Software Engineering Division of the MIT Instrumentation Laboratory—and one of the few women involved in the Apollo program—was instrumental in the development of a storage medium for software using an innovative method called rope memory, in which data was physically woven together by threading wires through magnetic cores. That same laboratory was by then employing microchips to accelerate development of Apollo guidance systems.

148 Back then, however: "Computers in Spaceflight: The NASA Experience," NASA, March 1, 1988.

150 If all went well: Wingo's payload was to be loaded onto Intuitive Machines' second flight.

## CHAPTER 26: Lunar Descent

156 The ground below: Low Sun angles, meanwhile, also posed novel challenges for charging solar panels, while the positioning of the lunar south pole relative to Earth could at times compromise real-time communications.

157 India's Chandrayaan-3: "NASA's LRO Observes Chandrayaan-3 Landing Site," Sept. 5, 2023.

160 Chatter at the party: Loren Grush, "On-the-Fly Software Patch Saved Intuitive Machines' Moon Mission," Bloomberg News, Feb. 23, 2024.

160 "did not like being": Steve Altemus interviewed by author.

163 Some antennas: Because of *Odie*'s orientation, ILOA's cameras were prevented from using their imaging camera to study astronomical bodies. As such, they were ultimately prevented from sharing that data with their Chinese counterparts.

163 "Today for the first time": In early March 2025, Intuitive Machines'

second lunar lander had targeted the southernmost landing site of any spacecraft in history. Roughly a hundred miles from the Moon's south pole, the flat-topped peak of a lunar mountain known as Mons Mouton spans nearly sixty miles and stands some twenty thousand feet above the surrounding terrain. And yet by most accounts, the daring mission to land there had the feeling of inevitability. A landing party had been orchestrated at Houston's Lone Star Flight Museum, adjacent to Ellington Airport, where the company housed its lunar production and operations center. Meanwhile, as barbecue was served, the cowboy hats and boots were out in a show of force. "This week, Texas owns space" was a common refrain. Another Texas space company, Firefly Aerospace, had early that week landed on an ancient basaltic plain, named Mare Crisium, located on the northeast quadrant of the lunar near side, having also worked under a comparative bargain NASA contract, just a shade over $100 million. Meanwhile, a separate Tokyo-based company called ispace had its second lander, Resilience, also on its way to the lunar surface, with subsequent plans to deploy AI-powered, "evolvable swarm robots that can share intelligence to settle in unknown and unexplored areas" of the Moon's surface in missions beginning in 2028. The three craft seemed to add fuel to those calling for the makings of a more viable lunar economy. Yet *Athena,* Intuitive Machines' newest vehicle, was bound for a decidedly different type of terrain. The lunar south pole, though more extreme, would be the locale of what many hoped would be the first human settlement. On board, it carried with it a series of NASA and commercial payloads, which included a rocket-powered drone, rovers, a drill to hunt for water, and equipment designed to test the makings of a 4G LTE lunar cell phone network.

But it was not to be. At least not this time.

While *Athena* appeared to experience virtually none of the hair-raising adventures of her old brother, *Odie,* en route to the Moon, the spacecraft still ultimately ended up on its side. In fact, it was declared dead less than twenty-four hours after landing—a mission that was supposed to last ten days—having fallen into a lunar crater, though not before transmitting images of its orientation and triggering a few experiments. Before any of that had been made public, however, Intuitive Machines' president, Steve Altemus, walked up to a microphone that rested atop a platform that overlooked the festivities and delivered a speech in a voice that sounded audibly deflated. "Well, hello, everybody," he began. "We had such a tremendous vehicle flying to the Moon," describing the fifteen minutes of powered descent as a veritable "nail-biter." During landing, the craft had employed machine-learning systems to avoid a series of boulders and ended up touching down several hundred meters from the intended site. "Noisy" data from a laser altimeter might have been linked to the faulty landing inside that crater, where extreme cold and the direction of the Sun prevented the craft from recharging batteries and operating as intended. As shares in Intuitive Machines plummeted, NASA's deputy associate administrator Joel Kearns said he was "disappointed in the outcome of the IM-2 mission," but also said the agency remained committed "to supporting our commercial vendors as they navigate the very difficult task of landing and operating on the Moon."

In a statement Steve later sent out, he addressed *Athena*'s final moments as the craft powered down. "Unexpectedly, the lander woke up one last time," with the craft transmitting "names of every Intuitive Machines team member who made her mission possible."

Still, that cratered landing raised real questions about the feasibility of touching down in such a harsh and unpredictable environment. Somehow human-rated landers were slated to go there so that they might forge the beginnings of a permanent settlement to then springboard out across the cosmos, just as the drumbeat of China's growing ambitions grew louder. Team members said they were "reloading," just as other companies looked to intensify their own efforts and NASA itself was embroiled in the throes of both structural and cultural change.

### CHAPTER 27: The SpaceX Effect

164 a National Security Space Launch contract: National Security Space Launch contracts service the U.S. Space Force and Missile Defense Agency.

164 It was a remarkable pace: Chengxin Zhang, "China May Need to Adjust Its Approach Toward SpaceX," *SpaceNews,* June 24, 2024.

165 As the two men waited: These quotations, and subsequent commentary derived from conversations with Lee Rosen and Frank Tybor, could not be independently confirmed with Elon Musk or representatives of SpaceX.

165 " 'Well, parachutes don't really' ": Atmospheric density of Mars is less than 1 percent of that on Earth, which means a parachute alone will not sufficiently slow down a descending craft. And yet NASA's Ingenuity Mars Helicopter nonetheless completed seventy-two flights starting April 19, 2021, proving that propellered flight is possible in the extremely thin Martian atmosphere. For more information, see NASA's complete write-up on the mission: Ingenuity Mars Helicopter, NASA, science.nasa.gov.

167 The more technical term: Bryan Palaszewski, "Entry, Descent, and Landing with Propulsive Deceleration: Supersonic Retropropulsion Wind Tunnel Testing and Shock Phenomena," NASA, Dec. 2012.

168 "Building many rockets": Elon Musk (@elonmusk), X, Feb. 20, 2020, 6:31 p.m.

168 "It didn't put a hole in the ship": The landing followed Blue Origin's New Shepard vertical touchdown in West Texas, which—in contrast to the Falcon 9—was a suborbital vehicle, meant as a prototype for future orbital launches. "Now safely tucked away at our launch site in West Texas is the rarest of beasts, a used rocket," Bezos said after the vehicle's engines reignited and eased the vehicle back onto the pad. But the differences were stark, not just because of the heights achieved.

170 The two-stage rocket: A modified Starship would also be slated for NASA's Artemis Human Landing System on the Moon.

170 a man named Vojtech Holub: Development of the technology had begun as early as 1969, when the cosmonauts Georgi Shonin and Valeri Kubasov first applied the welding technique off Earth, using a more rudimentary, though at the time highly experimental, remote-controlled tool called the Vulkan. Aboard the Soviet Soyuz 6 mission, three different types of welding experiments were conducted inside the craft: electron beam; low-pressure compressed arc, which is commonly used in fusing metals like aluminum and magnesium; and arc with a consumable electrode, which employs a molten metal bath.

### CHAPTER 28: The Art of War

172 Over the decades: Between AD 386 and AD 581, the region further fractured into competing northern and southern territories.

172 "read the stars": Patrick Hanan, "War Is Heaven," *New York Times,* Jan. 17, 1993.

173 The garrison had been spared: Ironically, years later during the Song dynasty, Kongming lanterns became symbols of blessings, celebration, and hope.

173 Incomplete information, misinformation: Carl von Clausewitz, *On War,* ed. and trans. Michael Howard and Peter Paret (Princeton University Press, 1976).

173 "into the fabric of modern": Steven L. Basham, "The Critical Role of Space in Modern Warfare and the Imperative of Joint Space Capabilities in Europe," U.S. European Command, Feb. 9, 2024.

174 the United States accounting for: Novaspace, *Government Space Programs: A Comprehensive Overview of Government Space Strategies, Activities, and Budgets Until 2033,* 24th ed., Dec. 2024.

174 President Trump's signature domestic policy: *Space Warfighting: A Framework for Planners,* United States Space Force, March 2025.

174 space was an active battleground: Federal clients seemed to increasingly lean on commercial partnerships to devise new architectures that might support defenses against threats in orbit. In fact, that shift had become abundantly clear only a few years earlier, just hours before the Russian military invasion of Ukraine. In 2022, Russian hackers launched a series of cyberattacks aimed at spreading disinformation and targeting Ukrainian financial institutions and communications infrastructure, which included satellite communications. Moscow expected rapid success and seemed initially intent on sowing confusion and hindering defenses, while simultaneously preserving the physical infrastructure, as its troops were deployed along Ukrainian borders. The moves, however, were not only expected but also revealed in near–real time by a commercial sector working in tandem with U.S. authorities. A Colorado-based company called Maxar Technologies, which says it provides the majority of commercially acquired geospatial intelligence for U.S. security efforts, snapped imagery of Russian equipment and soldiers building around Ukraine, including Crimea, and at an airfield in Belarus, which U.S. policy makers revealed as evidence of a pending Russia invasion.

175 "60 percent annually": "What Is Digital-Twin Technology?," McKinsey & Company, Aug. 26, 2024.

176 "to deliver effects": Deborah Lee James, Ryan McCarthy, and Michael E. White, "How the Nation Can Make Fielding Hypersonic Capabilities a National Priority," *SpaceNews,* March 6, 2025.

176 In fact, Russia, China, North Korea: In 2025, a review of scaling those technologies began in earnest to satisfy President Trump's call for a missile defense program akin to Israel's Iron Dome, designed to intercept short-range rockets, as well as shells and mortars. By contrast, America's renamed Golden Dome would require geographic coverage on orders of magnitude larger, and one that could also be employed in hypersonic defense. To fulfill the president's request, Booz Allen Hamilton in March 2025 proposed a defense constellation of some two thousand AI-powered satellites that would utilize a "multi-shot capability to defeat missile launches during the boost-ascent and early mid-course phases of flight." The system, called Brilliant Swarms, could apparently be employed against hypersonic boost-glide vehicles, as well as aerial drones and conventional cruise missiles, at a proposed price tag of $25 billion; it was just one of several proposals the White House had begun fielding.

And yet to realize such an ambitious defense measure (if that's indeed possible over America's 3.8 million square miles), enhanced monitoring, surveillance systems, and war-gaming capabilities were being considered of paramount importance.

177 "We'll measure more": The promise of such technologies is what, in part, prompted the U.S. Department of Energy in January 2025 to announce a $71 million initiative in high-energy physics—something that Regina Rameika, the associate director of the agency's Office of High Energy Physics, described as a path that could "[open] up new ways for us to understand and explore the universe."

The plans called for funding a variety of projects that included new experimental platforms for quantum technologies, such as entanglements—a phenomenon whereby particles are linked even when separated by large distances—as well as control of quantum states to more precisely track forces of gravity. The idea now was to find "like-minded people (and companies) that want to leverage these technologies," Rima Kasia Oueid, DOE senior commercialization executive at the Office of Technology Transitions, explained.

### CHAPTER 29: Turning Point

179 "Our job was to collect": Greg Hadley, "Saltzman: China's ASAT Test Was 'Pivot Point' in Space Operations," *Air and Space Forces Magazine,* Jan. 13, 2023.

179 Beijing had rolled out: Xichang Satellite Launch Center was then mainly responsible for the launching of the country's geosynchronous satellites.

180 "In that moment": Testimony of General B. Chance Saltzman, Chief of Space Operations, United States Space Force, Submitted to the U.S.-China Economic and Security Review Commission Hearing on China's Ambitions in Space, April 3, 2025, 10:00 a.m.

181 "in the interests of": Roscosmos, Telegram, March 17, 2024, t.me/roscosmos_gk/13597.

### CHAPTER 30: Starfish Prime

182 "It looked like noon": An account published in the Hawaii-based *Hilo Tribune-Herald* following the blast. Brian Gutierrez, "Why the U.S. Once Set Off a Nuclear Bomb in Space," *National Geographic,* July 15, 2021.

182 "It looked as though": Gutierrez, "Why the U.S. Once Set Off a Nuclear Bomb in Space."

182 Carried atop: "The first such failure due to total dose was the TELSTAR satellite, which was launched a day after STARFISH," according to a

subsequent NASA report, which also pointed to comparable Soviet high-altitude tests that same year over Semipalatinsk in Kazakhstan. "It was estimated that the spacecraft experienced a total dose from the weapon test 100 times larger than expected."

182 also destroyed roughly one-third: Liz Boatman, "Sixty Years After, Physicists Model Electromagnetic Pulse of a Once-Secret Nuclear Test," *APS News,* Nov. 10, 2022; E. G. Stassinopoulos, "The STARFISH Exo-Atmospheric, High-Altitude Nuclear Weapons Test," NASA, April 22, 2015.

183 "by several orders of magnitude": Stassinopoulos, "STARFISH Exo-Atmospheric, High-Altitude Nuclear Weapons Test."

183 thousands of satellites: Unshin Lee Harpley, "DOD Official Confirms Russia Is Developing an 'Indiscriminate' Space Nuke," *Air & Space Forces Magazine,* May 2, 2024.

183 even timing services: "What Is an Atomic Clock?," NASA, June 19, 2019. Atomic clocks mounted on GPS satellites measure the vibrations of atoms to measure time, and even then need twice-daily updates to correct for what NASA describes as "the clocks' natural drift." That time is then delivered to Earth as a gold standard known as Coordinated Universal Time, or UTC.

183 an attack on GPS: According to flight tracker data maps, tens of thousands of jamming and spoofing attacks are lobbed against commercial airlines every year, especially across eastern Europe and the Middle East. For clarity, jamming is a distortion or loss of signal when more powerful signals effectively drown out the intended ones, which has grown more prevalent with drones' capacity to interfere with GPS signals, while spoofing effectively tricks receivers into reporting that a vehicle, for instance, is somewhere it's not, sometimes by hundreds if not thousands of miles. The dangers in these instances can be profound, and include everything from navigational errors and elevated collision risks to a general loss of a pilot's situational awareness.

183 A widespread GPS outage: One of the advantages of the U.S. system, however, is its fragmentation, whereby individual businesses and localities often look to create their own redundancies that can make a coordinated national attack more difficult. China, by contrast, has taken a more centralized approach, and yet has been busy not only constructing its own alternatives but also implementing redundancy measures at the national level.

184 "The NASA Kennedy team": Leonard David, "Uncontrolled Reentry of Space Debris Poses a Real and Growing Threat," *SpaceNews,* June 5, 2024.

### CHAPTER 31: Space Junk

186 And there are more than 500,000: "NASA's Efforts to Mitigate the Risks Posed by Orbital Debris," NASA Office of Inspector General, Jan. 27, 2021.

187 That year, in February: 2009 Iridium-Cosmos Collision Fact Sheet, Secure World Foundation, Nov. 10, 2010; Anatoly Zak, "Strela Military Communications Satellite Family," Russian Space Web, Nov. 2, 2023.

Iridium 33 supported L-band mobile communications as a part of a broader "string of pearls" satellite constellation. The retired Russian satellite Cosmos 2251 was designed to record, store, and communicate data, useful in remote locales like Siberia, or in overseas embassies where highly classified information required secure communications.

187 When the two smashed into: 2009 Iridium-Cosmos Collision Fact Sheet.

187 "a very large, very energetic": Phillip Anz-Meador and J.-C. Liou, "Analysis and Consequences of the Iridium 33–Cosmos 2251 Collision," NASA.

188 Clean Space Initiative: Launched in 2012, the initiative identified four key areas of focus: (1) design and development of standards for spacecraft that help to reduce debris from launch to deorbit; (2) mitigation, which includes debris removal, collision avoidance, and satellite deorbiting; (3) remediation, which includes specific missions to capture derelict satellites and other objects; and (4) sustainable space logistics, which again focused on disposal. A year later, ESA research presented in Germany predicted collisions twice a decade based on the amount of debris that was then in orbit.

188 But, noted Heiner Klinkrad: Maria Sheahan, "Space Junk Needs to Be Removed from Earth's Orbit: ESA," Reuters, April 25, 2013.

188 In 2023, it nearly: Andrew Jones, "Close Call! 2 Huge Pieces of Space Debris Had a Near-Miss in Earth Orbit," Space.com, Sept. 21, 2023.

188 A year later, NASA's: George Dvorsky, "2 Satellites Almost Collided Above Earth," *Gizmodo,* March 1, 2024.

188 "too close for comfort": Leo Labs, X, Feb. 28, 2024, 9:46 a.m.

188 And as recently as October 2024: Two months earlier, in August 2024, a Chinese Long March 6A rocket, carrying a trove of communications satellites, shed more than fifty pieces of debris into low Earth orbit, unleashing new metal shards circling the planet.

188 The problem was clear: Space Environment Statistics, ESA, Environment Statistics, May 5, 2025.

189 That team had: Musk, who supported Trump during his presidential

campaign and donated more than $290 million to the 2024 election, according to federal filings, persuaded President Trump to put him in charge of the cost-cutting effort, just as SpaceX was emerging as one of America's largest contractors, having secured some $3.8 billion in federal contracts before Trump began his second term. By then, SpaceX had become the world's preeminent space company, while also assembling the planet's largest satellite constellation in Starlink, having produced roughly $8 billion in revenue in 2024 alone. The Pentagon, meanwhile, had grown especially reliant on both its internet services and the SpaceX Falcon 9 and Falcon Heavy rockets to deliver their own payloads into orbit, with senior leadership even discussing the potential for utilizing the company's superheavy Starship as a means of rapidly deploying supplies and troops around the globe.

190 "well below the elevation": David Shepardson, "DISH Gets First-Ever Space Debris Fine over EchoStar-7," Reuters, Oct. 2, 2023.

190 Photon pressure, ablation techniques, and lasers: Zero Debris Charter, ESA.

190 Debris collection efforts: In 1983, for instance, President Ronald Reagan unveiled a program called the Strategic Defense Initiative (SDI), which many later came to know as his Star Wars program, designed to protect the United States from a ballistic missile attack. Critics worried that the system upset the long-standing military doctrine of mutual assured destruction, in which nuclear deterrence relied on the notion that a launch would result in the annihilation of both the launcher and the recipient of the attack. The proposal, to critics, seemed costly and unrealistic. The Massachusetts senator Ted Kennedy referred to it as one of the president's "reckless 'Star Wars' schemes," taking a page from the 1977 film created by George Lucas. The name apparently stuck. But the system itself looked to directed-energy, particle beam, and other ground- and space-based weapons systems that many were concerned would contribute to the militarization of space. It wouldn't be the first time. Later, in the early '00s, the U.S. Air Force reportedly considered a proposal known as Project Thor, which would have put large tungsten rods into low Earth orbit. Those rods could then theoretically deliver strikes at targets on Earth. Tungsten, with its exceptionally high melting point and thermal conductivity, which allows for favorable heat dissipation during reentry, would essentially keep the rod intact as it struck like a meteor. It, like SDI, would be controversial, again about those broader militarization concerns.

190 For instance, when Starship's: Y. V. Yasyukevich et al., "Supersonic Waves Generated by the 18 November 2023 Starship Flight and Explosions:

Unexpected Northward Propagation and a Man-Made Non-Chemical Depletion," *Geophysical Research Letters* 51, no. 16 (2024); Davide Castelvecchi, "Huge SpaceX Rocket Explosion Shredded the Upper Atmosphere," *Nature,* Aug. 30, 2024.

190 Releases of aluminum oxide nanoparticles: José P. Ferreira et al., "Potential Ozone Depletion from Satellite Demise During Atmospheric Reentry in the Era of Mega-Constellations," *Geophysical Research Letters* 51, no. 11 (2024).

191 So its depletion can increase: European Commission, FAQ: Climate Action.

191 "peppered with particles containing": "NOAA Scientists Link Exotic Metal Particles in the Upper Atmosphere to Rockets, Satellites," NOAA Research, Oct. 13, 2023.

191 The findings, published: Daniel M. Murphy et al., "Metals from Spacecraft Reentry in Stratospheric Aerosol Particles," *PNAS,* Oct. 16, 2023.

192 "For a child": Kevin Loria, "Why Rocket Fuel Has Been Contaminating Our Food and Water for Years," *Consumer Reports,* Aug. 7, 2024.

192 Excess perchlorate exposure: "Perchlorate: Health Effects and Technologies for Its Removal from Water Resources," National Library of Medicine, April 14, 2014.

193 "A mishap investigation": Robert Z. Pearlman, "FAA Investigating SpaceX Starship Flight 8 Explosion That Disrupted Commercial Flights," Space.com, March 7, 2025.

### CHAPTER 32: Italian Entrepreneurs in the Red Desert

197 focusing on the impacts of prolonged microgravity: While in microgravity, blood and other fluids travel up from the legs and abdomen toward the heart and head, which can cause both swelling and a drop in total quantity of that fluid in the heart and blood vessels. That loss of blood in the heart, along with the atrophy of the heart as a muscle, can have effects tantamount to advanced aging that pose potential cardiac problems, and can actually alter the shape of the heart.

197 But it was Luca: "Every time we launch, we always have a server on board," Luca told me, as he detailed the growing importance of "our own cloud computing network in orbit." D-Orbit was focused on everything from reduced latency in communications between spacecraft and ground stations, data processing and storage, and the optimizing of bandwidth with machine-learning algorithms.

197 Having founded D-Orbit: By 2024, D-Orbit had raised more than $109 million in Series C funding, one of the largest funding rounds a European space company had ever enjoyed. More than half a dozen

major companies were now vying for their own slice of that market, with those like Spaceflight, Exolaunch, Momentus, Northrop Grumman, and Astroscale all in direct competition. But it was D-Orbit that in 2022 was valued at more than $1.28 billion, with plans to expand into satellite servicing and space cloud computing.

198 a company that devised solutions: "D-Orbit to Go Public in U.S. via $1.28 Billion SPAC Deal," Reuters, Jan. 27, 2022.

198 Nuclear-fueled propulsion: The so-called "tyranny of the rocket equation" tends to get in the way. In simple terms, it's essentially just a numbers game: To reach farther and farther distances, a rocket requires exponentially increasing amounts of fuel. In turn, that fuel makes the vehicle heavier and thus demands even more fuel, which in turn puts severe limitations on payload. That equation changes significantly if launched from the Moon, which has little atmosphere and one-sixth the gravity of Earth, and would thus allow vehicles to achieve comparable missions with far less propellant. But on Earth, capacity becomes but a fraction of the overall mass of the vehicle, which can create real problems for future Martian settlers in need of lots of cargo, including life support systems, food, water, power sources, and medical supplies, as well as communications, transportation, and thermal insulation systems, not to mention olive oil.

198 Median temperatures: Temperatures range from the 70s°F (20s°C) to -225°F (-153°C).

200 Crops would also need: Efforts to design "pioneer plants" to survive Martian conditions are under way, which could generate oxygen, food, and medicine, and soil conditions could then be improved by human waste. Those like Dr. Amy Grunden have spearheaded efforts to employ gene splicing, using genes from microbes that reside in the blistering waters emanating from deep-sea vents, inserting them in tobacco cells to infuse the plant with a degree of extreme weather tolerance as a prototype for future Martian plant life.

200 In a peer-reviewed study: M. I. Dobynde et al., "Beating 1 Sievert: Optimal Radiation Shielding of Astronauts on a Mission to Mars," *Space Weather* 19, no. 9 (2021).

200 "We are going to have to learn": "Why 'Living Off the Land' Will Be Crucial to the Dawning Era of Space Exploration," *Wired.*

201 "Here, we consider the more radical goal": Kevin M. Cannon and Daniel T. Britt, "Feeding One Million People on Mars," *New Space* 7, no. 4 (2019).

201 For argument's sake, if humans: It was January 20, 2025. The U.S. Capitol Rotunda filled with lawmakers and dignitaries as President Donald

Trump again stood at the lectern for another presidential inauguration. Some beamed. Others grimaced. Many wore painted smiles. But for the space community and their supporters in Congress, a lingering uncertainty seemed to permeate the sidelines of that event. Certainly, a lot had occurred during the president's first term. His administration had revived the National Space Council, established the U.S. Space Force, supported commercial partnerships, and pushed for a lunar return, all within the international framework of Artemis. The second round was presumed by many to be just as active. More than a month and a half before Inauguration Day, Jared Isaacman—a billionaire entrepreneur, pilot, and commercial astronaut—had already been nominated for the top job at NASA, a nomination that would subsequently be withdrawn. Elon Musk, meanwhile, who had spent millions supporting Donald Trump during the election, would find himself alternating between valued ally and contentious adversary to the administration. But at least at the outset, he seemed to have the president's ear. And on that day in January, Musk—who had long put the Red Planet in his sights—was all smiles as the newly elected president delivered his inaugural address.

"We will pursue our manifest destiny into the stars," said President Trump, "launching American astronauts to plant the Stars and Stripes on the planet Mars."

By then, Musk—who had laid an aggressive timetable for human missions to Mars by as early as 2029—could be seen giving the thumbs-up. He called the Moon a "distraction." By now, of course, space policy watchers were well acquainted with such shifts between administrations. Recall, for instance, how President Obama canceled the Bush administration's Constellation Moon program in 2010, redirecting NASA toward an asteroid-sampling mission instead. These were stark reminders of the often-fickle—and to some maddening—nature of American space policy. It was, however, more unusual that an administration shifts its own course, as those within the second Trump team appeared to be considering.

"The current legal imperative is that we continue with SLS [Space Launch System]," Michelle L. D. Hanlon, executive director of the Center for Air and Space Law at the University of Mississippi, told me, referring to Title 51 of the U.S. Code, which governs national and commercial space programs. In 2018, Congress added language that codified NASA's exploration goals into law: The administrator "shall establish a program to develop a sustained human presence in cis-lunar space or on the moon," which it called "a stepping-stone to future exploration of Mars and other destinations." SLS and Orion were mentioned by name in subsequent passages, all of which seemingly compelled NASA to incorporate those rockets and capsules into its plans for again landing astronauts on the Moon.

By that reading, any attempt to send humans would have to wait until at least after 2027—the current target for the first crewed landing of the Artemis program. A presidential or executive order cannot override Title 51, given it is federal law, enacted by Congress, which maintains constitutional power over funding and legal authority in space policy. And while a president can shift direction within the boundaries of the law, he or she could not legally override or ignore laws like Title 51 without congressional action, especially when such explicit Moon-then-Mars language is written into law. Yet with an administration that would soon take a more expansive perspective of executive power, and with a Congress that would soon afford the White House a measure of discretion in spending matters, deeper questions were forming about who precisely was steering the nation's space policy, and if SLS's clear lack of sustainability warranted enough of a rationale for a change in direction under the premise of an emerging national security threat posed by China's coming dominance in space.

Certainly commercial interests were playing a larger role, with Musk and SpaceX exerting an unprecedented level of influence within American space policy. What then might a corporate-driven future on the Red Planet start to resemble?

This is when a look at the archives can prove useful, especially when tracing back to the very roots of the United States itself.

### CHAPTER 33: Governing Mars

203 "refusing any Earth-based authority": Antonino Salmeri, "Op-ed: No, Mars Is Not a Free Planet, No Matter What SpaceX Says," *SpaceNews,* Dec. 5, 2020.

206 "[Chinese workers] won't even": Wilfred Chan, "Elon Musk Praises Chinese Workers for 'Burning the 3am Oil'—Here's What That Really Looks Like," *Guardian,* May 12, 2022.

### CHAPTER 34: Where Is Everybody?

208 advanced alien civilizations: In September 2025, a sample collected by NASA's Perseverance Mars rover in the Jezero Crater contained potential biosignatures of ancient microbial life, according to a paper published in the journal *Nature.* "This finding by *Perseverance,* launched under President Trump in his first term, is the closest we have ever come to discovering life on Mars," said acting NASA administrator

Sean Duffy in separate comments, just months before Jared Issacman was renominated to lead the agency. "NASA's commitment to conducting Gold Standard Science will continue as we pursue our goal of putting American boots on Mars' rocky soil." J. A. Hurowitz et al., "Redox-Driven Mineral and Organic Associations in Jezero Crater, Mars," *Nature* 645 (2025): 332–340.

208 "Fermi grasped that any civilization": "The Fermi Paradox," SETI Institute.

210 "all technologically advanced species": Marek Abramowicz et al., "A Galactic Centre Gravitational-Wave Messenger," *Scientific Reports,* April 27, 2020.

211 "If aliens visit us": "Aliens," *Into the Universe with Stephen Hawking,* directed by Martin Williams, aired April 25, 2010, on Discovery.

211 "misses a crucial possibility": Abraham Loeb, "Space Archaeology," Harvard University, Nov. 8, 2019.

212 While astronomers speculated: "Nobel Winners Changed Our Understanding with Exoplanet Discovery," NASA, Oct. 8, 2019.

## CHAPTER 35: The Search

214 It also uncovered evidence: Europa is roughly one-fourth the diameter of Earth, and yet NASA estimates its ocean could contain as much as twice the water of Earth.

214 A never-before-seen molecule: "Webb Makes First Detection of Crucial Carbon Molecule," NASA, June 23, 2023.

215 the metal-rich asteroid Psyche: Mainly comprising iron and nickel, as well as platinum and gold, it drew headlines after its value was estimated at upwards of $100,000 quadrillion, or $100,000,000,000,000,000,000. That absurdly large figure, of course, ignores some basic rules of supply and demand, as well as the more fundamental question of how one might actually retrieve an asteroid that large and return it to Earth.

A $1.4 billion solar-electric-powered NASA spacecraft is expected to survey the asteroid by August 2029.

215 a California-based company: The James Webb Space Telescope's evaluations of the curious space rock drew headlines, in part because of speculations over what might eventually constitute an emerging space mining industry. In fact, in August 2024, AstroForge announced that it had raised $40 million in a bid to "help unlock the near-limitless resources in space," totaling roughly $55 million to date. The funds, in actuality, were to be used for the company's third mission, Vestri, on board Intuitive Machines' IM-3 lunar lander, whereby the craft would detach from the IM craft, travel to a nearby asteroid, and then land

on it as an effective mining surveyor. Still, the effort was expected to be a first of its kind for a commercial space company. With the U.S. Congress having allowed private companies rights to the resources they mine in space, companies like AstroForge and Interlune, the latter having been built at least in part by former Blue Origin executives, were now collaborating with NASA and other space companies to develop viable new markets in space, particularly when it came to asteroids.

215 "Will we actually land": Jacqueline Feldscher, "A Post-Mission Debrief with AstroForge's CEO," Payload, April 16, 2025.

216 "one-kilometer-diameter asteroid": Jonathan O'Callaghan, "Earth's 1st Asteroid Mining Prospector Heads to the Launchpad," *New York Times,* Feb. 23, 2025.

### CHAPTER 36: Defending Earth

217 "tens to hundreds of nuclear": "Asteroid Apophis: Will It Hit Earth? Your Questions Answered," Planetary Society.

218 "Our uncertainties are": Paul Wiegert, "On the Sensitivity of Apophis' 2029 Earth Approach to Small Asteroid Impacts," *Planetary Science Journal* 8, no. 5 (Aug. 26, 2024).

219 In the air above Chelyabinsk: Phil Black, Boriana Milanova, and Laura Smith-Spark, "Russian Meteor Blast Injures at Least 1,000 People, Authorities Say," CNN, Feb. 15, 2013.

219 A University of Arizona–led: ESA also offered up a $68 million contract to the Italian aerospace company OHB Italia to study Apophis.

220 To change that: Near-Earth Object Surveyor, NASA.

220 "This is a problem we know": Jeff Foust, "NEO Surveyor Launch Delayed Despite Funding Boost," *SpaceNews,* Jan. 26, 2023.

### CHAPTER 37: Impact

223 The technology was supposed: The mission also carried with it an Italian Space Agency–devised CubeSat, which could capture images of impact and the resulting debris cloud.

223 "tiny maneuvering error": Johns Hopkins University Applied Physics Laboratory, DART interactive explainer.

223 had endeavored to reach: NASA's Near Earth Asteroid Rendezvous mission touched down on the asteroid Eros for approximately $150 million, not adjusted for inflation, on February 12, 2001.

225 A subsequent study also raised: Harry Baker, "Fallout from NASA's Asteroid-Smashing DART Mission Could Hit Earth—Potentially Triggering 1st Human-Caused Meteor Shower," Live Science, Aug. 27, 2024.

225 Those bits would be expected: Ed Reynolds—an APL veteran of thirty-

seven years—who had served as DART's mission project manager and would go on to be named one of *Time* magazine's most influential people for his work at DART, had also hoped that the mission would serve as a proving ground for an experimental ionized propulsion system, which essentially fires propellant gas at a considerably higher velocity than the more traditional chemical thrusters. DART came equipped with hydrazine propellant for maneuvering and attitude control, as well as xenon to power this ion propulsion demonstration. The technology effectively employed a gridded ion engine called a NEXT-C (or NASA's Evolutionary Xenon Thruster–Commercial), which had been developed by the agency's Glenn Research Center and Aerojet Rocketdyne, a subsidiary of the U.S. defense company L3Harris Technologies, which builds rocket, hypersonic, and electric propulsive systems. The system was meant to be an improvement on ion propulsion systems flown during NASA's previous Dawn and Deep Space 1 missions. And yet there had been a problem.

"We had designed [the system for electrical] currents . . . that could go through the chassis . . . of up to 25 amps," Ed explained. "We thought the two-hour test was all nominal. And then it was later that we discovered we had something that we could not explain in our power system," with modeling showing currents as high as 100 amps. "It was not something that we had designed our spacecraft to withstand," he added of the fourfold increase. And so the decision was made not to fire DART's demonstration of NEXT-C, fearing it could damage the craft and undermine the broader mission. "We didn't want to risk that." Still, Ed seemed convinced about the future of ion propulsion and its growing use in deep space missions. Simultaneously, a host of new technologies were also coming online. "We're always thinking about what our next mission is," he added.

### CHAPTER 38: Alpha Centauri Awaits

229 Without knowing it, his mom: Moms have a way of doing things like that. My own mom studied aerospace engineering during the 1960s in the hopes of joining those Apollo astronauts, though instead became a critical care nurse who scored "an amazing job at the 'Space Station,' a.k.a. the intensive care unit." Her son never quite devised plans for the next star system over, however.

229 After speeding away from: Estimates suggest that it won't be until November 18, 2026, that Voyager 1 will reach a full light-day's distance away from Earth.

229 Again, Alpha Centauri is at a distance: A light-year is the distance that light can travel in one year across the near vacuum of space.

230 Launched in 2018: Mara Johnson-Groh, "NASA's Parker Solar Probe Completes 23rd Close Approach to Sun," NASA, March 25, 2025.

230 They called it Proxima b: The Moon has no discernible forms of life, in part due to its lack of atmosphere and extreme climatic conditions.

230 "extraterrestrial technosignature": Sofia Z. Sheikh et al., "Analysis of the Breakthrough Listen Signal of Interest blc1 with a Technosignature Verification Framework," *Nature Astronomy* 5 (2021): 1153–62.

230 "undoubtedly one of the most": "What Was That Signal from Proxima Centauri?," SETI Institute, Oct. 26, 2021.

231 That shorter distance might: On Earth, our planet's system effectively forms a protective bubble, along with another recently discovered field known as Earth's ambipolar electric field, considered "as fundamental as Earth's gravity and magnetic fields," which was first hypothesized during the late 1950s and the 1960s.

231 solar flares: On May 1, 2019, the largest ever solar flare was recorded. Proxima Centauri, which is about one-eighth the size of the Sun, reportedly "went from normal to fourteen thousand times brighter when seen in ultraviolet wavelengths over the span of a few seconds," according to the astrophysicist Meredith MacGregor, an assistant professor at the Center for Astrophysics and Space Astronomy and the Department of Astrophysical and Planetary Sciences at the University of Colorado Boulder.

231 kind of atmosphere that sufficiently absorbs: In August 2024, NASA's Endurance mission confirmed the existence of the ambipolar field, which until that point had only been theorized. "Something had to be drawing these particles out of the atmosphere," noted Glyn Collinson, principal investigator of Endurance at NASA's Goddard Space Flight Center.

231 And given that uncertainty: Based on the analysis of the astrophysicist Paul Sutter. Paul Sutter, "Life on Proxima b Is Not Having a Good Time," Universe Today, Dec. 11, 2022.

231 in a computer modeling study: K. Garcia-Sage et al., "On the Magnetic Protection of the Atmosphere of Proxima Centauri b," *Astrophysical Journal Letters* 844, no. 1 (2017).

"The question is, how much of the atmosphere is lost, and how quickly does that process occur?" noted Ofer Cohen of the University of Massachusetts, Lowell, who coauthored the study. "If we estimate that time, we can calculate how long it takes the atmosphere to completely escape—and compare that to the planet's lifetime."

See also "An Earth-Like Atmosphere May Not Survive Proxima b's Orbit," NASA, July 31, 2017.

232 Under President Reagan: Melissa Healy, "'Star Wars' Enters Second Decade with No Hardware, $30-Billion Tab: Defense: Debate over Its Future Centers on Its Scope. The Objectives Continue to Change Drastically from Reagan's," *Los Angeles Times,* March 22, 1993.

232 an airborne infrared observatory: "SOFIA" stands for Stratospheric Observatory for Infrared Astronomy.

233 "billions of hidden": Kepler / K2, NASA.

233 The moons of Jupiter: Venus, though generally inhospitable, has temperatures that range between 30°F and 200°F in its upper atmosphere, which has fueled speculation that Venetian clouds could be an area where life could theoretically exist. Scientists using the James Clerk Maxwell Telescope in Hawaii and the Atacama Large Millimeter Array observatory in Chile made the discovery of the potential biosignature gas. But their findings are hardly conclusive. Under certain circumstances, phosphine can be produced through nonbiological processes. But the findings nonetheless prompted the New Zealand–based company Rocket Lab to fund a private mission to visit Earth's nearest planetary neighbor. Jane S. Greaves et al., "Phosphine Gas in the Cloud Decks of Venus," *Nature Astronomy* 5 (2021): 655–64.

234 "Earth is the cradle": Konstantin Tsiolkovsky, "Earth Is the Cradle of Humanity but One Cannot Live in the Cradle Forever: Taken from a Letter Written by Konstantin Tsiolkovsky in 1911," European Space Agency.

234 In 2005, Cosmos 1: "Russians Say Solar-Sail Vehicle Was Lost," Associated Press, June 21, 2005.

234 the first solar sail in space: Technically, solar sailing in space had unexpectedly occurred much earlier, in 1974, when NASA's Mariner 10 faced a problem of insufficient propellant needed to adjust course in between the orbits of Venus and Mercury. Faced with few options, the teams adjusted the craft's solar panels in a manner that used photons from the Sun to correct course, thus enabling the first-ever close-up images of Mercury.

In 1976, Carl Sagan—a co-founder of the Planetary Society—promoted the idea of light sails on *The Tonight Show* with Johnny Carson, originally calling for the craft to rendezvous with Halley's Comet.

### CHAPTER 39: Interstellar Voyagers

236 Once a spacecraft: That doesn't mean there aren't factors that can slow it down over time. Gases and dust pepper the cosmos and create a degree of drag. Gravitational forces from nearby planets and asteroids also exert

their own influence, as do the radiation pressures from other stars that can blunt a craft's momentum over time.

237 "Space travel as we know it": Yuri Milner's parents reportedly named him after the Soviet cosmonaut and first man in space, Yuri Gagarin.

238 "generate an optical signal": Thomas Eubanks, "Swarming Proxima Centauri: Coherent Picospacecraft Swarms over Interstellar Distances," NASA, Jan. 4, 2024.

### CHAPTER 40: Party in the Valley

240 "Yuri [Milner] invited us": This response is based on Pete Worden's recollection. Neither Will Marshall nor Yuri Milner was available for comment.

241 Even Microsoft's co-founder: Stoke Space, "Bill Gates' Breakthrough Energy Leads $65M Funding Round for Stoke Space's Reusable Rockets," press release, Dec. 2022.

241 "When you have abnormal behavior": "Dr. Vadim Smelyanskiy, Principal Scientist, Ames Research Center, Moffett Field, CA," *Tech Briefs,* March 27, 2014.

243 the concept was akin: An analogy initially offered up by Peter Riva.

243 And back then, Pete saw: At the time, in 2013, there was just one company (more or less) believed to be making a commercial quantum computer. D-Wave, based in Burnaby, Canada, had garnered significant attention (and skepticism) for its so-called adiabatic computer, thought to be especially adept at solving optimization problems.

243 "I think from the Google side": Qubits are considered dynamic subatomic particles. As such, they do not always interact in the same predictable ways, making error correction especially important. Google's Bristlecone, a 72-qubit quantum chip, was at the time considered a significant advancement in the field of quantum computing given its relative processing power and error correction.

244 "cautiously optimistic that quantum": "A Preview of Bristlecone, Google's New Quantum Processor," Google Research, March 5, 2018.

244 "In reaching this milestone": Frank Arute et al., "Quantum Supremacy Using a Programmable Superconducting Processor," *Nature* 574 (2019): 505–10.

244 "the ultimate thermonuclear device": Arthur Herman, "Q-Day Is Coming Sooner Than We Think," *Forbes,* Jan. 7, 2021.

244 "We are improving the precision": Jane Cai, "Does a Talent Crisis Threaten China's Quantum Ambitions? One Chinese Expert Thinks So," *South China Morning Post,* Oct. 24, 2024.

244 By 2024, Beijing's heavy investments: Owen Hughes, "China Creates Its Largest Ever Quantum Computing Chip—and It Could Be Key to Building the Nation's Own 'Quantum Cloud,'" Live Science, May 13, 2024.

245 "I'm sure we can arrange it": Schmidt could not be independently reached to confirm these statements. Attribution is therefore placed entirely on interviews with Worden and his recollection of events.

247 How would we know: Those laws initially included three principal guidelines for robotics: (1) "A robot may not injure a human being or, through inaction, allow a human being to come to harm"; (2) "A robot must obey orders given to it by human beings except where such orders would conflict with the First Law"; (3) "A robot must protect its own existence, as long as such protection does not conflict with the First or Second Law." He later included the "Zeroth Law," which expanded upon the first law in stating that "a robot may not harm humanity, or, by inaction, allow humanity to come to harm."

247 In fact, a growing body of research: "Study Supports Quantum Basis of Consciousness in the Brain," *Neuroscience News,* Sept. 6, 2024.

248 In fact, NASA commissioned: Jason Dunn, "Reconstituting Asteroids into Mechanical Automata," NASA, April 7, 2016.

### CHAPTER 41: Nuclear Rockets

250 A "pusher plate," affixed: "The advantage of the nuclear rocket of this kind over the chemical type lies paradoxically not so much in its potentially enormous power source, which is limited by chamber temperature T to much the same range as chemical motors, but in its ability to use hydrogen as propellant," the report noted. Subsequent iterations discarded the idea of a combustion chamber in favor of releasing the bombs directly behind the craft.

251 Almost simultaneously, Project Rover: "Nuclear Rockets," NASA.

251 And within just a few years: And as recently as 2023, the old proposals seemed to be gaining new life. That year, NASA and the Defense Advanced Research Projects Agency announced that Lockheed Martin and BWX Technologies would collaborate on designing different aspects of a $499 million propulsion system known as DRACO, or Demonstration Rocket for Agile Cislunar Operations, that could effectively halve the amount of time it would take to reach Mars. The DRACO system would use high-assay low-enriched uranium in a nuclear reactor, which could also be employed for the ultrafast maneuvering of satellites. And yet the fiscal 2026 budget didn't include any funding for DRACO, with it being effectively canceled.

251 By 1958, the United States: Kenneth Chang, "NASA Seeks a Nuclear-Powered Rocket to Get to Mars in Half the Time," *New York Times,* July 26, 2023.

253 Freeman Dyson had even suggested: For context, lighter nuclei fuse together and create heavier nuclei in stars, unlike fission, where atoms are split. Within the Sun, that process is known as proton-proton fusion, which is conversion of relatively small amounts of mass into enormous amounts of energy. It is a phenomenon that has long intrigued researchers interested in its potential application in space, especially given the process can be fueled by isotypes of the most abundant element in the known universe: hydrogen.

253 using fusion energy pulses: In fact, charged particles that emanate from the Sun's corona (often called solar wind) have for billions of years deposited an isotope found to be useful for nuclear fusion (as well as in the cooling of high-performance quantum computers) in the lunar regolith.

### CHAPTER 42: Harnessing the Power of the Sun

255 "the potential importance": Directorate of Intelligence, "The Soviet Magnetic Confinement Fusion Program: An International Future," CIA Special Collections, Release as Sanitized, 2000.

257 The concept was reportedly classified: Elena Bonner, "Sakharov Is Tokamak's Originator," *Physics Today* 58, no. 12 (2005).

258 the first so-called tokamak systems: The term "tokamak" comes from the Russian words *toroidalnaya kamera magnitnaya katushka* (or toroidal chamber magnetic coil).

259 In a bid to make it: Elizabeth Gibney, "ITER Delay: What It Means for Nuclear Fusion," *Nature,* July 8, 2024.

260 Such rockets could drastically: Titan, a moon larger than the planet Mercury, is one of the solar system's few places with a subsurface ocean and atmosphere. Effectively leaning on roughly thirteen years of Cassini data, the initial mission is expected to land in the equatorial Shangri-La dune fields and explore the region in a series of progressively longer flights of up to five miles. A multi-rotor NASA vehicle, akin to a large drone, will make the journeys, and is expected to visit the Selk impact crater to explore for evidence of liquid water and organics. The trips will "nearly double" the distance traveled to date by all the Mars rovers combined, according to an agency statement. "Titan is unlike any other place in the solar system, and Dragonfly is like no other mission," noted NASA's associate administrator for science Thomas Zurbuchen. "It's remarkable to think of this rotorcraft flying miles and miles across

the organic sand dunes of Saturn's largest moon, exploring the processes that shape this extraordinary environment. Dragonfly will visit a world filled with a wide variety of organic compounds, which are the building blocks of life and could teach us about the origin of life itself." The mission is part of NASA's New Frontiers program, which also includes missions to Pluto and the Kuiper Belt.

260 Such a system could: Many of those blueprints were decades old. In fact, during the 1970s, researchers at the British Interplanetary Society drummed up a proposal for a two-stage vehicle that employed an inertial confinement engine, using electron beams to compress the vehicle's fuel, heat it, and trigger fusion. The so-called Project Daedalus was a design study meant to review the feasibility of interstellar travel "using current or 'near' future technology." The idea was revived decades later with Project Icarus, a theoretical engineering design study that borrowed much of the former's structure, though it ended in 2019 without a clear plan forward. Propellant would essentially compress and heat a magnetized plasma to a level suitable for fusion, according to the NASA scientist John Slough, who coauthored a 2012 paper, "Nuclear Propulsion Through Direct Conversion of Fusion Energy: The Fusion Driven Rocket." Once in orbit, a magnetic nozzle would then propel thermal energy into high-velocity exhaust, which—according to Slough and others—could be "realized with little extrapolation from currently existing technology." The resulting exhaust could conceivably propel the craft at fractions of light speed.

260 a private fusion market: "The Global Fusion Industry in 2023: Fusion Companies Survey," Fusion Industry Association.

260 Another seven listed: "Global Fusion Industry in 2023."

260 "It's irresistible to": Aria Alamalhodaei, "Pulsar Fusion Wants to Use Nuclear Fusion to Make Interstellar Space Travel a Reality," *TechCrunch*, July 7, 2023.

260 For a fleeting moment: Matthew Sparkes, "UK Nuclear Fusion Reactor Sets New World Record for Energy Output," *New Scientist*, Feb. 8, 2024.

260 Efforts to shrink: It would not conduct its first operation until 2034—half a century after the joint project was first proposed. It would also take five years after that before the first major experiments were conducted, adding an additional $5 billion to the endeavor.

261 the potential to release: Dr. Mike LaPointe, "Antimatter Propulsion," NASA.

**CHAPTER 43: Ask a Fish About Water—a Visit to CERN**

262 It would have been enough: The name CERN is derived from the acronym for the French Conseil Européen pour la Recherche Nucléaire, or European Council for Nuclear Research.

262 "Physicists tell a parable": Brian Greene, "How the Higgs Boson Was Found," *Smithsonian Magazine,* July 2013.

263 After protons were accelerated: It was Higgs on which a physicist named Peter Onyisi had focused his work. With thin glasses and a slight stoop, presumably from years hunched over this vast subatomic world, Peter seemed to embody the very notion of a particle physicist. He became an associate professor at the University of Texas at Austin and was also part of the CERN team that made the discovery. "I was a weird kid," he told me when we met, harking back to his small hometown in Nigeria. The privileged son of an ambassador, with critical access to a British Council library set in a leafy residential neighborhood of the capital, he'd slip away after school to wander its stacks, often thumbing through science books, including the works of the astrophysicist John Gribbin, among them his book *In Search of the Big Bang*. First published in 1986, the work examined the origins of the universe, its expansion, and whether it would "one day recollapse into a mirror image of the Big Bang." Peter remembers reading it intently as his father played John Denver albums in the adjacent room. "There's me reading about supersymmetry while he's playing 'Rocky Mountain High' on the cassette deck," he said. It was then when Peter first encountered this concept of the Higgs field. And by the time he arrived at CERN, scientists had power to search for it using their recently built Large Hadron Collider, constructed just around the time Peter was earning his PhD from Cornell. "I was lucky, time-wise," he said. "Of course, if the one in Texas had been built, we would have found the Higgs a long time ago."

263 It was an ambitious project: "Reverse Brain Drain? Exploring Trends Among Chinese Scientists in the U.S.," Stanford Center on China's Economy and Institutions, July 15, 2024; Yu Xie et al., "Caught in the Crossfire: Fears of Chinese-American Scientists," *PNAS,* June 27, 2023.

264 And by the spring of 2025: Gerrit De Vynck and Danielle Abril, "Tech Companies Are Telling Immigrant Employees on Visas Not to Leave the U.S.," *Washington Post,* March 31, 2025.

265 Indeed, the goal was: Ling Xin, " 'Two Sessions' 2024: China's Construction of the World's Largest Particle Collider May Start in 2027," *South China Morning Post,* March 8, 2024.

### CHAPTER 44: The Antimatter Factory

267 Just a single gram of the stuff: In Switzerland at CERN, currently the world's largest particle physics laboratory, those efforts had already involved the exploration of antimatter, particles that effectively hold the same mass as ordinary matter particles but are composed of the opposite charge. A founder of the field of quantum mechanics, the Nobel Prize–winning British physicist Paul Dirac identified a problem when combining quantum theory and special relativity as it related to the behavior of electrons traveling at close to the speed of light. In his groundbreaking 1928 relativistic wave equation—also known as the Dirac equation—which helped predict the behavior of electrons in ways consistent with Einstein's theory of relativity, Dirac surmised that every particle must have a corresponding antiparticle, which was effectively identical, only with an opposing charge. At a basic level, particles comprise electrons, protons, and neutrons, whereas antimatter is made up of positrons, antiprotons, and antineutrons. The concept, however, should equate to equal amounts of each particle following the Big Bang's creation of the universe. So far as we can tell, at least thus far, it does not. In fact, the amount of antimatter in the known universe is found in such infinitesimal amounts that it raises a natural question: Where did it all go? Could there be entire antimatter galaxies out there to effectively offset that asymmetry? That imbalance is among the reasons why even though the Standard Model of particle physics is generally considered an extremely powerful model, it is also considered incomplete. Scientists at facilities like CERN were intent on finding clues that might offer more clarity. And yet while the research conducted here had gained renown for speeding up and colliding particles, which had uncovered the Higgs, the sprawling complex also had the capacity to find—and trap—antimatter.

268 Its potential in space propulsion: Antimatter was later discovered by the physicist Carl Anderson while studying cosmic rays at the California Institute of Technology. In 1933, he and the physicist Seth Henry Neddermeyer employed gamma rays to generate positrons.

268 Three years later: On June 28, 1919, Germany and the Allied powers signed a peace agreement known as the Treaty of Versailles, which officially ended World War I and established the conditions of peace, which included reparations and a loss of German territory that involved roughly 10 percent of its previous population. Hitler later considered the treaty a national humiliation. In a speech delivered in Munich on April 13, 1923, he described the treaty's intention as an effort "to

bring twenty million Germans to their deaths, and to ruin the German nation." On March 7, 1936, he sent troops back into the Rhineland, a region that bordered France. The action was thought to embolden further aggression after no real countermeasures were taken. Three years later, German forces invaded Poland, prompting Great Britain and France to declare war on Germany, and thus igniting World War II.

269 In a paper published in 2017: Espen Gaarder Haug, "The Ultimate Limits of the Relativistic Rocket Equation: The Planck Photon Rocket," *Acta Astronautica* 136 (July 2017): 144–47.

269 the potential use of neutrinos: Bruce Dorminey, "Neutrinos to Give High-Frequency Traders the Millisecond Edge," *Forbes,* April 30, 2012.

## CHAPTER 45: Warp Drives

271 The sci-fi series: The character of Jean-Luc Picard, captain of the USS *Enterprise,* was played by Patrick Stewart, who coincidentally studied only an hour's drive across the channel at the Bristol Old Vic Theatre School.

272 That essentially constitutes: Dark energy, which does not emit, absorb, or reflect light and yet is predicted through interactions with normal matter, is thought to make up about 68 percent of the universe. Its properties remain largely unknown.

It was an idea developed in earnest by the Russian theoretical physicists Alexei Starobinsky and Andrei Linde, as well as the American theoretical physicist and cosmologist Alan Guth. The idea is that the universe expanded violently in the first few fractions of a second, stretching the very fabric of space-time much faster than light could travel, then slowed, and has since reaccelerated, presumably at least in part due to the presence of dark energy.

274 "The Warp Drive": Miguel Alcubierre, "The Warp Drive: Hyper-Fast Travel Within General Relativity," *Classical and Quantum Gravity* 11, no. 5 (1994).

275 In this framework: The nature of quantum mechanics does, however, challenge classical intuitions about the nature of information across space and time—particularly in relation to phenomena like entanglements, an occurrence in which two or more particles are linked in such a way that the state of one particle directly correlates with the state of another, no matter the distance between them.

## CHAPTER 46: Casimir

277 "powerful intellect and exceptional memory": C. M. Hargreaves, "Hendrik Brugt Gerhard Casimir, Knight of the Order of the Nederlandse

Leeuw, Commander in the Order of Orange Nassau, 15 July 1909–4 May 2000," *Biographical Memoirs of Fellows of the Royal Society* 50 (2004): 39–45.

279 Masahiro Hotta, a theoretical physicist: Masahiro Hotta, "Quantum Measurement Information as a Key to Energy Extraction from Local Vacuums," *Physical Review D* 78, no. 4 (2008).

Two decades later, in a first-of-its-kind breakthrough that is somewhat aligned with Hotta's vision, researchers at Northwestern University announced that they had successfully teleported a quantum state of light across eighteen miles of fiber-optic cable, something nobody thought possible, explained Prem Kumar, professor of electrical and computer engineering and director of the Center for Photonic Communication and Computing, who led the study.

280 "In principle, it may be possible": Richard Obousy and Aram Saharian, "Casimir Energy and the Possibility of Higher Dimensional Manipulation," Oct. 24, 2018.

281 His paper, titled: Harold "Sonny" White, "Warp Field Mechanics 101," NASA.

281 the paper seemed to harken: The reactor, nicknamed Chicago Pile, first achieved a sustained nuclear reaction on December 2, 1942.

284 "the conversation about warp drives": "New Study Achieves Breakthrough in Warp Drive Design," Business Wire, May 7, 2024.

## CHAPTER 47: Adaptation

288 Today, even a brief survey: As we enter this planet's sixth era of mass extinction in service of an ever-expanding global economy, biodiversity is disappearing at a staggering pace. The rate of species extinction "is already tens to hundreds of times higher than it has been, on average, over the last 10 million years," according to UN estimates, derived from a report compiled by 145 expert authors from fifty countries, with inputs from another 310 contributing authors, in what the global body says is a first-of-its-kind accounting of "the relationship between economic development pathways and their impacts on nature."

With some 60 percent of the world's oceans now being "maximally sustainably fished," an atmospheric rise in $CO_2$ by 50 percent since the onset of the industrial era, and a more than 100 percent growth of urban areas since 1992, human populations have become ever more reliant on expansions of food, energy, and consumer goods that commonly come at the expense of global sustainability goals. It's something Sandra Díaz, an Argentine ecologist who cochaired the UN assessment, described as an approaching "breaking point."

288 As rapidly advancing technologies: The trappings of artificial intelligence stretch back to 1950, when Alan Turing posed the simple question "Can machines think?" Yet perhaps it goes even further to 1666, when Gottfried Leibniz argued that ideas were little more than combinations of a finite set of concepts.

# Index

*(Page references in italics refer to illustrations.)*

## A NOTE ABOUT THE AUTHOR

DAVID ARIOSTO is a journalist, space industry analyst, and co-host of the podcast *Space Minds* on *SpaceNews*. A columnist at *Aerospace America, SpaceNews,* and *Noema,* as well as visiting scholar at Arizona State University's Interplanetary Initiative, he also serves as a strategic adviser to space companies around the globe. His debut book, *This Is Cuba: An American Journalist Under Castro's Shadow,* was published in 2018 and was highly acclaimed by *The Washington Post* and *USA Today*. He lives in Arizona with his daughter, Rose, and their dog, Cali.

## A NOTE ON THE TYPE

This book was set in Adobe Garamond. Designed for the Adobe Corporation by Robert Slimbach, the fonts are based on types first cut by Claude Garamond (ca. 1480–1561). It is to him that we owe the letter we now know as "old style." He gave to his letters an elegance and feeling of movement that won him an immediate reputation and the patronage of Francis I of France.

Composed by North Market Street Graphics
Lancaster, Pennsylvania

Designed by Betty Lew